The Grip of Gravity
The Quest to Understand the Laws of Motion and Gravitation

Gravity is the most enigmatic of all known forces of nature. It controls everything, from ocean tides to the expansion of the Universe. The search for the laws of motion and gravitation started over two thousand years ago. The reader is taken on an exciting journey through the subsequent centuries, identifying the blind alleys, profound insights and flashes of inspiration that have punctuated this search. Despite the fantastic progress that has been made, the true nature of gravity is still a mystery and the book attempts to show how the current developments in string theory (perhaps the 'Theory of Everything') may lead to a new and radical interpretation of gravity. This book describes the fundamental concepts and developments, and the experiments, both performed and planned, to increase our understanding of gravity and the natural phenomena in which gravity is the principal player.

PRABHAKAR GONDHALEKAR is an astrophysicist with major research interest in the interstellar medium and active galactic nuclei. He was formerly head of the Space Astronomy Group at the Rutherford Appleton Laboratory.

PRABHAKAR GONDHALEKAR

The Grip of Gravity

The Quest to Understand the Laws of Motion and Gravitation

PUBLISHED BY THE PRESS SYNDICATE OF THE UNIVERSITY OF CAMBRIDGE
The Pitt Building, Trumpington Street, Cambridge, United Kingdom

CAMBRIDGE UNIVERSITY PRESS
The Edinburgh Building, Cambridge CB2 2RU, UK
40 West 20th Street, New York, NY 10011–4211, USA
10 Stamford Road, Oakleigh, VIC 3166, Australia
Ruiz de Alarcón 13, 28014 Madrid, Spain
Dock House, The Waterfront, Cape Town 8001, South Africa

http://www.cambridge.org

© Cambridge University Press 2001

This book is in copyright. Subject to statutory exception
and to the provisions of relevant collective licensing agreements,
no reproduction of any part may take place without
the written permission of Cambridge University Press.

First published 2001

Printed in the United Kingdom at the University Press, Cambridge

Typeface Trump Mediaeval 9.5/15pt *System* QuarkXPress™ [SE]

A catalogue record for this book is available from the British Library

Library of Congress cataloguing in publication data

Gondhalekar, Prabhakar, 1942–
  The grip of gravity: the quest to understand the laws of motion and gravitation /
  Prabhakar Gondhalekar.
    p. cm.
  Includes bibliographical references and index.
  ISBN 0 521 80316 0
  1. Gravitation – History. 2. Motion – History. I. Title.
  QC178.G56 2001
  531'.14–dc21   00-065149

ISBN 0 521 80316 0 hardback

Contents

Preface   vii

Aristotle   1
    Greek science   2
    Alexandrian science   12
    Islamic science   19

Kepler   25
    The Copernican revolution   28
    The professional astronomer   35
    Marriage of mathematics and observations   42

Galileo   51
    Mechanical philosophy   55
    Birth of experimental physics   61

Newton   77
    Origins of dynamics   81
    Laws of motion   91
    Universal gravitation   98
    Newton's universe   101
    Laws of Kepler and Newton   102
    Conserved quantities   104
    Action   108
    After the *Principia*   111

Einstein   119
    Speed of light   124
    What is a straight line?   126
    Doubts about Newtonian mechanics   131

vi CONTENTS

    Theory of special relativity 142
    Theory of general relativity 155
    Choice between Newton and Einstein 170
    Alternatives to general relativity 173
    Consequences of general relativity 176

Dicke 189
    Universal constant of gravitation G 191
    Inverse square law 203
    Gravitational redshift 207
    Deflection of light by gravitational field 211
    The 'fourth test' or the time delay of radio waves 218
    Equivalence Principle 223
    Gravitomagnetism 233
    Gravitational waves 238

Hubble & Eddington 247
    Figure of planets 247
    Tides 249
    Ice ages 251
    The universe – an expanding fire-ball 254
    Galaxies, quasars and clusters of galaxies 275
    Dark matter 286
    The birth, life and death of stars 295
    Neutron stars 308
    Black holes 311
    Planets and planetary systems 320

Planck 325

Chronology 342

Notes 344

Suggestions for further reading 348

Index 351

Preface

Our world is ruled by two sets of laws: the laws of gravity and the laws of quantum mechanics. The laws of gravity describe the large structures in the universe such as the Earth, the solar system, stars, galaxies and the universe itself. These laws allow us to predict the path and motion of spacecraft and asteroids and also the evolution of the universe. The laws of quantum mechanics, on the other hand, describe the very small structures such as molecules, atoms and subatomic particles. They enable us to understand the three subatomic forces, lasers, CD players and nuclear weapons. One of the great puzzles of the twentieth century is that these two sets of laws, each employing a different set of mathematics and each making astonishingly accurate predictions in its own regime, should be so profoundly different and incompatible.

Quantum mechanics is a child of the twentieth century. Its origins can be traced back to the year 1900, when Planck proposed the particle nature of electromagnetic radiation to explain the black-body spectrum. The character of the laws of motion and the laws of gravity, on the other hand, has unfolded over a considerably longer period. Today, the concepts of mass, force and gravity are very familiar, but they are also deeply mysterious and are intimately linked to our understanding of motion. Historically motion was perhaps the first natural phenomenon to be investigated scientifically. Over two thousand years ago, the Greek philosopher Aristotle made the first attempt to make the concept of motion more precise. Unfortunately he coupled this with his doubtful astronomical views and separated the motion of the celestial bodies from the free fall of objects on Earth. This separation impeded the understanding of the laws of motion and the development of ideas of universal gravitation until the middle ages. The study of motion or mechanics was also the first 'science' to be developed in the modern period starting with Isaac Newton in the seventeenth century. The internal beauty and elegance of Newton's work and its early success in accounting in quantitative detail for the motion of the Moon, and of other planets, had enormous influence on philosophical thought and provided impetus for systematic development of science in the twentieth century. By introducing the concept of universal gravitation Newton swept aside the separation of celestial and terrestrial motions which had been assumed for the previous two thousand years. The concepts of space, time and relativity enter naturally into the study of motion. Newton assumed absolute and independent space and time without actually using these concepts in the application of his mechanical principles. Two centuries after Newton the question of absolute motion arose again, this time in connection with electrodynamics. It was Albert Einstein's great genius to accept, finally, that there was no such thing as absolute and independent space and time. This simple but revolutionary admission led Einstein to his now famous equation $E = mc^2$.

Gravity has a strong grip on human imagination, and Newton and Einstein dominate the development of gravitational theory. Newton's classical theory held sway for two hundred years. At the beginning of the twentieth century it was realised that Newton's theory does not describe the motion of bodies in a strong gravitational field or bodies that move close to the speed of light. In his theory of general relativity Einstein reinterpreted the concept of gravity. He showed that gravity could be described in terms of the geometry of space-time. Einstein's field theory of gravitation predicts only small departures from Newtonian theory except in circumstances of extreme gravitational fields or high speeds. But the major significance of Einstein's theory is its radical conceptual departure from classical theory and its implications for the future development of scientific thought.

Normally, the laws of gravitation and quantum mechanics operate completely independently, but in situations of extremely high gravitational attraction like the surface of a black hole or on extremely small length scales like the point of origin of the universe, the two sets of laws come together. In these situations the two sets of laws act together in ways which we still do not understand. It now seems likely that the theory of general relativity leads to an approximation of the true nature of gravity; the true form will only be found in the synthesis of general relativity and quantum theory or the theory of quantum gravity. It is astonishing that three hundred years after Newton and one hundred years after Einstein the quest for gravity still continues.

In this book I have traced the gradual unfolding of our understanding of the laws of motion and universal gravitation and the associated concepts of space, time and relativity. This unfolding has taken a long time (and is still continuing) and many fascinating personalities have been involved in the process. Experimental verification has played an essential part in our understanding of these laws and a number of challenging experiments are planned to deepen that understanding. Gravity has fashioned our universe and the story of gravity would not be complete without a brief review of the astronomical processes in which gravity is a major player.

This book is intended for both nonspecialists and students of science. For science students and teachers, I hope this book will expose part of the foundation of modern physics. For nonspecialists, I hope that by describing the evolution of this major theme in science, I have given a feel for the long road that has, we believe, led us to the brink of the "theory of everything".

Many people have helped me to write this book: Barry Kellett, David Giaretta and George Hanoun helped with various aspects of word processing and text preparation. David Pike read the first draft of the book with great care and generously bandaged the wounds I had inflicted on the English language. Francis Everitt read the manuscript closely, and I am grateful for his detailed comments on a number of aspects of this book. I particularly appreciate his comments on the historical details and the latest technical information, on the proposed space experiments to test the theory of gravitation, which he provided. Jayant Narlikar also offered valuable comments on a number of topics in the book. Awinash Gondhalekar reviewed the near-complete manuscript: inclusion of his comments and observations on all aspects of this book has considerably enhanced this presentation. I would like to thank Lindsay Nightingale, who read the final version of this book with great care; her comments and questions were very helpful in 'making the science clear' at a number of points in the book. Lastly I would like to thank the editorial and production teams of Cambridge University Press. Any errors and omissions are, of course, entirely my responsibility.

Finally, I would like to thank my wife Jane for her unwavering support.

Prabhakar M. Gondhalekar

1 Aristotle

The Bronze Age (2000 BC–1000 BC) was a period of major advances and changes in the riverine cultures of Mesopotamia, Egypt, India and China. Exceptional developments in irrigation and agriculture had led to the establishment of large urban civilisations in which arts and sciences were encouraged and patronised. Efficient tax collection and revenue management had freed funds to support a class of people who could devote their time to study, observation and contemplation. The alphabet and numbers were formalised and the practice of keeping records, both civil and military, was well established. Concepts of space and measurement and concerns with heavenly bodies and physical structures led to the development of arithmetic and geometry.

The development of predictive and exact sciences followed from the study of motions of the Sun, Moon and the five visible planets. The periodicity of the motion of these bodies was utilised to establish a quantitative measure of time, and the correlation between the rising and setting of groups of stars and seasons was developed into a calendar, which we use today (in slightly modified form). The systematic study of the motion of the Sun and Moon by the Mesopotamian priest-astronomers enabled them to identify the cause of eclipses and also to predict future eclipses. The discipline of record keeping had been extended to note the occurrences of unusual astronomical events and irregularities in the movements of planets. The belief-systems developed by these cultures defined man's position in the cosmos and his relationship with nature. The ethical system they established led to the development of the concepts of 'cause' and 'effect'. These concepts were later extended to philosophical and scientific speculations and form the basis of modern scientific thought. However, by 600 BC these cultures were undergoing a crisis; the spiritual legacy of the Bronze Age was being questioned. A number of great thinkers of the world were alive at this time: Buddha in India, Pythagoras and the early Greek philosophers, the Old Testament prophets in Israel and Confucius and Lao Tzu in China. Fundamental questions were being asked about the nature and purpose of life. It was also a time of great speculation about the natural world and there seems to have been a recognition of the existence of 'natural laws' in the universe. This was a period of change and it has been called the Axis Age.

GREEK SCIENCE

In Europe the stirrings of scientific inquiry started among the Ionian Greeks of Asia Minor. In the Western cultures today it is generally believed that Greeks and Greeks alone 'invented' science. This Eurocentric (or Greco-centric) sentiment indicates a fundamental lack of understanding of the nature of scientific inquiry. Scientific inquiry is a process and by its very nature it is progressive. New ideas and discoveries are built on foundations and structures erected by others at earlier times. The Ionians and Greeks generally were in close contact with older cultures further to the East and South, and also in Asia. It is certain that Greek science in its origin was dependent on knowledge and traditions that came from ancient civilisations of Egypt and Mesopotamia. On this the Greeks and in particular the Greek historian Herodotos have insisted, and modern discoveries confirm it. We will never know the full extent of 'borrowing' by the Greeks from other cultures. But to the Greeks we owe the formal and conscious development of science as a discipline and the synthesis of observations or empirical relations to expose fundamental unifying principles.

The Golden Age of Greek science was during the fifth and fourth centuries BC. This was the period of the famous teacher–pupil sequence Socrates, Plato and Aristotle. They laid the foundation of natural philosophy which was to dominate the Islamic and Christian cultures for centuries to come. They believed in the existence of absolute or universal truths, which could be discovered or deduced by a system of logic. The founder of this 'system of thought' was Socrates (470–399 BC), whose overwhelming preoccupation was with 'conduct'. He firmly believed that everything was created and carefully controlled by a Supreme Being. He also stressed the importance of the soul and its persistence after death. The body was regarded as a temporary habitat for the soul (there are strong similarities here with the Indian/Buddhist view of the body and soul). Socrates left no written record but according his pupil, Plato, he was well versed in geometry and astronomy. He believed that astronomy was useful for determining the day of the year or the month, but all speculations about the motion or orbits of planets were regarded as a complete waste of time. In 399 BC Socrates was accused of 'impiety' and was put to death by drinking hemlock. The triumph of the Socratic doctrine held back for a while the development of Greek science and physical philosophy but it also led to the emergence of two giants of science in the fourth-century BC, Plato and Aristotle.

Plato (427–367 BC) was a student of Socrates and like him he was concerned with ethical motives. He believed that true morality was as immutable and objective as geometry, and discernible by use of reason. His school, the Academy, persisted for many centuries and was chiefly preoccupied with metaphysical discussions. Plato drew a distinction between reality and appearance and between knowledge and opinion. To him the everyday world of senses was worthless because it was a product of opinion. True knowledge was in the mind and consisted of pure ideal form. By implication the human body itself was a shadow; only the soul was real. This became the central tenet of the Neoplatonist Christians in the Middle Ages. But Plato was also an accomplished mathematician ('Let none who has not learnt geometry enter here' was inscribed over the entrance to the Academy) and had Pythagorean teachers. Many of Plato's thoughts have a mathematical guise. Plato appealed to other sciences to exhibit the certitude and exactness demonstrated by mathematics. He also had high regard for astronomy because in his opinion the heavenly bodies, in their motion, demonstrate the perfect geometric forms favoured by Pythagoreans. However, Plato wanted to explain the universe, not simply describe it, and his emphasis was on the theoretical aspects of astronomy rather than observations. Plato regarded the irregularities of planetary motion to be inconsistent with his view of the perfect universe. These irregularities had, in his opinion, to be explained in terms of simple circular motions. Plato accordingly set his students to seek out the rules by which the motion of planets could be reduced to simple circles and spheres. This task was to preoccupy astronomers for the next two thousand years.

One of the first students of Plato to distinguish himself in science, and in astronomy in particular, was Eudoxus (409–356 BC). He was an observational astronomer and not a theorist, as preferred by Plato. He accurately determined the length of the solar year to be 365 days and 6 hours (this was already known to the Egyptians). But his most influential contribution was in following up Plato's contention that the orbits of the heavenly bodies *must* be perfect circles. Eudoxus proposed that the heavenly bodies move on a series of concentric spheres with the Earth (which was assumed to be a sphere) at the centre. Each planetary sphere rotates around an axis, which is attached to a larger sphere that rotates around another axis. The secondary sphere was succeeded by a tertiary and a quaternary sphere, as required to explain the annual and the retrograde motion of a planet. For the Sun and Moon, Eudoxus found that three spheres were sufficient. To explain the movements of other planets, four spheres were required. The motion of the fixed stars could be explained with just one sphere. In all, 27 spheres were required to explain the movements of all known bodies in the sky. More spheres were added as further irregularities of the heavenly bodies were discovered. Another pupil of Plato, Heracleides (388–315 BC), first suggested that the Earth completed one full rotation on its axis in 24 hours. He also correctly attributed the motion of Mercury and Venus to their revolution round the Sun. It is not known whether he realised that this was also true of the other planets. It was almost another 1800 years before the ideas of Heracleides were generally accepted.

Aristotle (384–322 BC) was born in Stagira in Macedonia. He arrived at Plato's Academy in 367 BC when he was 17 years old. On Plato's death in 347 BC he moved to the Aegean island of Lasbos because of the increasing anti-Macedonian sentiment in Athens. In 343 BC he returned to Macedonia to tutor the young Alexander, then 13 years old. In 336 BC Alexander embarked on his career of conquest and Aristotle returned to public teaching in Athens. There he owned a garden called the Lyceum where he established a school, later called the Peripatetic (Greek for 'walking around'), where he lectured and taught. He and his associates and students carried out research on scientific and philosophical topics. Under his direction the school also produced a monumental account of the constitutions of Greek city-states. In 323 BC Alexander died and Athens once again became hostile to Macedonia and Macedonians. Aristotle was accused of impiety (the charge that had been levelled against Socrates) and he was forced to flee north to Chalcis, a Macedonian stronghold, leaving the Lyceum in the hands of his colleagues. He is said to have remarked that he would 'not allow the Athenians to sin twice against philosophy'. Removed and isolated from the cultural stimulation of Athens and his school, Aristotle died a lonely man in 322 BC at the age of 62.

Aristotle was driven by a desire for knowledge and understanding in every possible realm. His works cover every topic from (A)stronomy to (Z)oology and they are teeming with detailed observations about the natural world and also abstract speculations. He believed, and this was his unique gift to the world, that the universe was not controlled by blind chance or magic, but by a set of rational laws, which could be discovered, analysed and catalogued to guide human behaviour. His output of work was prodigious, but sadly only about a quarter of it has survived. His earliest work was on biological subjects, which was probably written during his stay on Lesbos. Most of his later work was probably written during his second stay in Athens.

Aristotle's biological work is based on first-hand observations of living things, and it is this research which established him as a man of science. In his *History of Animals* he describes a scheme for the classification of all living things, from plants to humans. This was a grand synthesis and remained the ultimate authority for many centuries after his death. In his second Athenian phase he turned to the investigation of physical and astronomical problems and set forth a general view of the universe. But in stark contrast to his biological studies his investigations of physical and astronomical problems were devoid of observations or personal knowledge. His physical and astronomical conceptions had profound influence on the centuries that followed, but his biological work was neglected and eventually forgotten, to be rediscovered only in recent times. His writings on natural philosophy suggest that he was attempting to synthesise, in a general scheme, the structure of the material world, not unlike his earlier biological synthesis. Aristotle was looking for an order in both the physical and the biological worlds, and to him these two were related.

Aristotle's world-view, which was to dominate the European view of nature for two thousand years, was based on a common-sense picture of the universe. This can be summarised thus:

Matter is continuous

There was considerable speculation on the nature of matter even in pre-Socratic Greece. In the fifth century BC, Democritus (470–400 BC) and his followers the Epicures had postulated an atomic nature of matter. According to this theory all matter is composed of solid *atoms* and the space or *voids* between them. To Democritus the voids were as much a primary reality as the atoms themselves. The atoms were considered eternal, indivisible and invisibly small. They were also considered incompressible and homogeneous and differed only in form, size and arrangement. Movement or rearrangement of atoms produced the qualities that distinguish things. Democritus and his followers showed little tendency to extend their scientific ideas further and the atomic theory was practically forgotten until the eighteenth century.

Opposed to the atomists were the continuists, among them the fifth-century scientist and philosopher Anaxagoras (488–428 BC) and later Aristotle (and also Socrates and Plato). The continuists believed that all matter was composed of a primordial stuff called *hyle*. Aristotle's stature was such that his views acquired dogmatic authority in the medieval Christian church and the atomic view of matter became particularly abhorrent.

In Aristotle's universe, below the sphere of the Moon

> All matter is made of four fundamental 'elements', earth, water, air and fire that interact and are capable of transforming into one another. Each element is characterised in turn by four 'qualities', heat, cold, dryness and moisture and these occur in pairs.

This concept was not original to Aristotle and is of considerably more ancient origin. It appears to be based on the observation that everything in nature and everyday life has a fourfold division: four seasons, four directions, the four ages of man. This concept fits in well with Jewish, Christian and Islamic thought and became part of orthodox medieval theology. The concept of four qualities was not challenged until the seventeenth and eighteenth centuries.

In this universe

> Planets, stars and the Moon are made of a different kind of matter, a fifth element – *quintessence*. The natural movement of the fifth element is circular and it is eternal. The heavenly bodies are attached to crystal spheres that rotate with a uniform circular motion around an axis passing through the stationary Earth which is at the centre. Each sphere is influenced by the spheres outside it.

The sphere of atmosphere surrounds the Earth and around that are respectively spheres of earthy exhalations, water, air and fire. These spheres are pure elements and are not accessible to humans. Beyond the sphere of elemental fire is a sphere of even more exotic substance, the *ether* (Greek: 'shining') or *quintessence*, which enters into the composition of the heavenly bodies. Beyond the sphere of ether are in succession the seven spheres of the Moon, Sun and the five planets then known and beyond these the sphere of the fixed stars. Finally, beyond all these spheres is the sphere whose divine harmony keeps all the other spheres in motion. Aristotelian teaching maintained that at creation the 'Prime Mover', God, had set the heavens in perfect and eternal circular motion. The crystalline spheres moved, he said rather obscurely, by 'aspiring' to the eternal unmoved activity of God. In order for this 'aspiration' to be possible he assigned a 'soul' to these spheres. Motion was communicated from the Prime Mover to the sphere inside it, and so on to the inner spheres. There was no such thing as empty space, as all space was filled with God's presence. Aristotle, like Plato, was influenced by Pythagorean concepts of 'perfect forms and figures' and in particular circles and spheres as being 'more perfect' than most. His world-view was, therefore, based on these concepts. The heavenly spheres he conceived were in agreement with the mathematical scheme of Eudoxus.

The cosmology of Aristotle was a product of the Greek anthropocentric world-view. The Greeks and, following them, the European Christians were obsessed with the notion that man was central to God's creation and to them the centrality of Earth in cosmology was self-evident. This belief prevailed until the time of Copernicus in the fifteenth century and elaborate schemes were developed both by the Greek and European thinkers to accommodate it. Also, in Aristotle's universe

> Circular motion is perfect and represents changeless, eternal order of the heavens. In contrast, motion in a straight line is confined to our changing and imperfect world.

The basis of this concept is that the heavenly bodies, which were set in motion by divine intervention, appear to be in circular motion, which therefore must be perfect and unaffected by external causes or agents. The four sub-lunar elements tend to move in a straight line: earth downward towards the centre of the universe, fire towards the extreme, and air and water towards intermediate places. Once in its natural place, each of the four elements remains at rest unless caused to move. The difference in the motion of the heavenly bodies and those on the Earth remained a puzzle till the end of seventeenth century when Newton provided a self-consistent explanation of both motions.

Aristotle differed sharply from medieval and modern thinkers in that he believed that

> The universe is finite in space, the outer limit being defined by the 'sphere of divine harmony', but infinite in time, there being neither creation nor destruction as a whole.

The questions of whether space and time are finite or infinite arise when one contemplates the universe at large. These questions have been considered for centuries but, even now, we cannot claim to have a definitive answer. Aristotle concluded that the material universe must be spatially finite. His reasoning was quite simple: if the stars extended to infinity then they could not perform a complete revolution around the Earth in twenty-four hours. Space itself must also be finite because it is only a receptacle for material bodies. Aristotle also asserted that the universe was temporally infinite, without beginning or end, since it is imperishable and can be neither created nor destroyed. These concepts became the backbone of the Christian Church in the Middle Ages. In the sixteenth century Giordano Bruno challenged this view. He asked a simple and obvious question: if there is an edge or a boundary to the universe, what is on the other side? For his curiosity (or impertinence!) Bruno was burned at the stake (see Chapter 2, Kepler).

Aristotle coined the word 'physics' from the Greek word *physis*, or nature, to designate the study of nature. He was perhaps the first scientist/philosopher to consider the concepts of motion, inertia and gravity. To him motion was fundamental to nature; he was to declare, 'To be ignorant of motion is to be ignorant of nature'. Aristotle's physics was accordingly the science of *natural* motion: that is, motion resulting spontaneously when a body is released from all constraints.

Heavy bodies had natural motion downwards, the natural motion of light bodies was upwards. He regarded the motion of the celestial bodies and the behaviour of the four elements as 'natural' motions; all other motions were 'violent motions'. To him the natural state of a body was to be stationary; nothing moves unless it is pushed. But his laws of motion were not based on observations or experiments: they were stated as being self-evident. He believed that:

> The application of 'force' (*dunamis*) displaces a body by a distance proportional to the time of application of the force.
>
> Objects in motion seek their natural place of equilibrium

Aristotle believed that a body moving at constant speed requires a force to act on it continuously and that force can only be applied by contact with the body – a proximate cause. He also believed that the four elements had an inherent force, which drove them to their natural place of rest. A solid body (made of the first element, earth), on being dropped, will fall vertically in a straight line to its natural place of rest, the centre of the universe (or Earth). A heavy body will fall faster than a light body and the speed of fall is proportional to the size of the body. Since Aristotle's physics was not concerned with forced motion, he gave no plausible explanation for the continued motion of a thrown object after it leaves the hand. Because he had postulated physical contact with a mover to account for any motion that was not natural, he suggested that the medium through which a body moves assists the motion of the body. The idea was that if a body is projected through air then the air that is displaced will rush round behind the body to provide the motive force for the body. Aristotle also had no concept of composition of motion or force; he argued that if a running man threw an object vertically up it would fall down and land behind the man. He had clearly conjectured this and not taken the trouble to observe it, a significant departure from the methodology he employed in his biological work. Aristotle's laws of motion could not be used to learn more about how bodies move. His physical science was also nonmathematical; it was qualitative and lacked the abstraction that is the power of modern physics. This was true of Greek science generally. In ancient Greece there was no tradition of experimental verification or inductive projection, now considered essential for any scientific inquiry. Aristotle's speculative views appeared rational because of their apparent agreement with 'common sense'. Because of this 'common-sense' nature, his views on motion and astronomy remained entrenched for the next two thousand years, impeding the understanding of the principles of motion, and precluding the emergence of ideas of universal gravitation. Galileo finally overthrew his views on motion during the Renaissance, by returning to the methodology employed by Aristotle in his biological studies – observations and experiments.

In Aristotle's writings we see a fundamental departure from the basic aim of Plato. Plato was concerned with the question 'Why' in his investigations of nature. Aristotle was concerned with the question 'How' in his work. Plato attempted to identify an underlying 'law' to explain the complexity of nature while Aristotle attempted to explain this complexity in as concise and consistent a manner as possible. This dual 'track' of scientific inquiry is followed today.

Aristotle's work covered huge areas of knowledge, his best scientific work being in biology and his worst in physics. His works were rediscovered and enthusiastically adopted by both Islamic and Christian scholars. Many Christian scholars accepted every word in his writings as eternal truth, as long as it did not contradict the Bible. The sixth century Greek/Egyptian philosopher and theologian John Philoponus identified Aristotle's 'Prime Mover' with the Christian personal god. This Christianisation of Aristotelian doctrine continued in the following centuries and ultimately was fused and reconciled with the Christian doctrine into a philosophical system known as Scholasticism. This became the official philosophy of the Roman Catholic Church.

Because of this emphasis on the Aristotelian world-view, some scientific discoveries in the Middle Ages and Renaissance were criticised by the Church simply because they were not mentioned by Aristotle. Aristotle's stature was such that he unwittingly created an intellectual *cul-de-sac* in which European and Islamic physical science was trapped for centuries. There can be no doubt that it was his authority that delayed the full development of such sciences as dynamics and astronomy. Almost two thousand years later the founding father of modern physics (and astronomy), Galileo, had to struggle and suffer to reject the doctrines of Aristotle before setting science on a firm foundation.

ALEXANDRIAN SCIENCE

By 300 BC, Athens had ceased to be a centre of scientific inquiry. After the first generation of Peripatetic scholars, Aristotle's school devoted itself to preserving and commenting on the writings of its founder. With the end of the Platonic and Aristotelian era there was also an end to philosophical attempts to define the universal character of nature. Science and philosophy parted company and the two proceeded along their own peculiar paths to achieve their limited objectives.

The torch of intellectual activity was handed over to Alexandria, a city in Egypt founded by Alexander of Macedonia. There, Ptolemy II founded a library that was to become a centre of intellectual activity for the next five hundred years. In Alexandria a synthesis of mathematics and astronomy (and science in general) was achieved, and quantitative astronomy with its powerful predictive capability was born. Also in Alexandria, mathematics assumed an independent position. This independence is emphasised by the famous *Elements of Geometry* of Euclid (330–260 BC). It is unlikely that all of *Elements* is Euclid's original work. Others had written elementary works on geometry before Euclid, but in the thirteen books of *Elements* Euclid presents the old knowledge and new work in terms of propositions (theorems, often mixed in with problems) and where necessary definitions, postulates and axioms. Euclid was very sophisticated in his treatment of questions of definition and proof and careful about stating the necessary and sufficient conditions of theorems. For example, his parallel axiom states that 'given a straight line and a point outside the line, it is possible to draw one and only one line through the point, which is parallel to the given line'. He defined 'parallel straight lines' as those which, 'if extended indefinitely in both directions, do not meet each other in either direction'. The concept of flat space is implicit in these propositions. In such space the sum of the angles of a triangle is 180 degrees. Euclid did not consider lines and figures on curved space. The concept of a universal flat space was to be challenged only in the early twentieth century, and this challenge revolutionised our concepts of motion and gravitation.

From the second century BC the synthesis of mathematics and science was taking place throughout the Mediterranean region. Archimedes (287–212 BC) of Syracuse in Sicily produced profound works on geometry, mechanics, hydrostatics and military engineering. He is credited with the exposition of the principle of levers. Archimedes did not invent levers, which in their various forms had been in use from remote antiquity. But Archimedes gave a formal and systematic description of them, a description which was susceptible to exact analysis. Archimedes also set down the fundamental principles of mechanics as rigorous geometric propositions. For example, equal weights at equal distances from the point of support (*fulcrum*) are in equilibrium. This led him eventually to the discovery of the centre of gravity of various geometric figures.

The story of Archimedes leaping from his bath crying 'Eureka! Eureka!' (I have found it! I have found it!) is well known and may well be apocryphal but what Archimedes had found was the concept of specific weights of bodies. The scientific aspects of this concept are described in his *On floating bodies*. Archimedes also made significant contributions to mathematics, perhaps the greatest being his method to measure the area of curved figures and surfaces. The method had been invented and used before him but Archimedes introduced the systematic use of the idea of 'limits'. Euclid discusses this idea in some detail in his twelfth book. The idea of limits is essential to 'calculus' as developed by Newton and Leibniz (see Chapter 4, Newton). In his method of proof Archimedes must be regarded as a modern scientist, but his physical insight did not extend beyond objects at rest and he did not consider objects in motion.

Mathematics and mathematical methods were also changing the character of astronomy. Eratosthenes (276–194 BC), the librarian at Alexandria, inaugurated the era of celestial measurements. He measured the circumference of Earth by measuring the angles cast by shadows of two *gnomons* (vertical rods) a known distance apart. He also measured the obliquity of the ecliptic, that is, the angle the circle of the zodiacal constellations makes with the celestial equator. About a hundred years later Hipparchus of Nicaea (190–120 BC), who worked at Rhodes, started a programme of accurate measurements of the positions of the fixed stars and compiled a star catalogue containing 850 stars. This was to be of profound importance to future astronomers. He also collected the observations of fixed stars made by Babylonian[1] and earlier Greek astronomers. When Hipparchus compared his observations with those obtained earlier he found that there were differences in the positions of the stars when measured from the same fixed point in the sky. This led him to establish that the tropical year (i.e. the period from the beginning of spring in one year to the beginning of spring next year) was shorter than the sidereal year (i.e. the time taken by Earth to revolve round the Sun once, as measured by its return to the same group of stars on the ecliptic) by about 20 minutes. Hipparchus also correctly surmised that the vernal (spring) equinox moves westwards by about 50 arcseconds every year (this phenomenon was also known to the ancient Babylonians). This westward drift of the vernal equinox suggested that the axis of Earth rotates in the direction of the apparent daily motion of the stars. The complete cycle of precession takes 26 000 years. The physical and mathematical rationale for this motion of Earth's axis was provided by Newton in the seventeenth century.

In 50 BC Egypt became a province of the Roman Empire, but by then most of Alexandria's intellectual achievements were behind it. The creative curiosity had given way to a considerable capacity for compilation and archiving. However, a few original sparks remained and notable among them was Claudius Ptolemy (127–145 AD), no relation to the Greek rulers of Egypt. Virtually nothing is known about his life, but he was an accomplished scholar and wrote extensively on geography, mathematics and astronomy. His great astronomical thesis has come down to us via the Arabs as the *Almagest* (an Anglicised version of the Arabic *Al-majestic*, which itself is a 'corruption' of Greek *'megale mathematics syntaxis'*, i.e. great mathematical composition). It consists of thirteen books, each dealing with certain astronomical concepts pertaining to stars and objects in the solar system. It is intended as a comprehensive mathematical presentation of the motion of the planets as seen in the sky and covers both orbits and distances of heavenly bodies. It is difficult to determine which parts of the *Almagest* are those of Ptolemy and which are those of Hipparchus. But Ptolemy did extend some of the work of Hipparchus; for example, the star catalogue of Hipparchus contains 850 stars, and Ptolemy expanded this to 1022 stars in his catalogue. He also gave an extensive description of astronomical instruments. Ptolemy's *Almagest* marks the zenith of Greek astronomical achievements. Ptolemy also extended the conclusions of Hipparchus to formulate his geocentric theory, now known as the Ptolemaic system. He built upon the conceptions of his predecessors: Aristotle for the notion of a finite spherical universe with a stationary Earth at the centre and Hipparchus for mathematical methods. He gave various arguments to 'prove' that in its position at the centre of the universe, the Earth must be immovable. He argued that since all bodies fall to the centre of the universe, as had been asserted by Aristotle, the Earth must be fixed at the centre of the universe, otherwise falling bodies would not be seen to drop towards the centre of the Earth. He also argued that if the Earth rotated on its axis once every 24 hours, a body thrown vertically upwards would not fall back to the same spot as it is seen to do. As a result of these arguments the geocentric system was dogmatically asserted in Western Christendom until the fifteenth century, when Copernicus finally challenged it.

Ptolemy accepted the Aristotelian order of the celestial objects – Earth (centre), Moon, Mercury, Venus, Sun, Mars, Jupiter and Saturn. He had also realised, like Hipparchus before him, that to preserve the supposed uniform circular motion of the heavenly bodies, yet accommodate their irregularities (like the retrograde motion of some planets[2]), it was necessary to assume either a system of deferents and epicycles or one of movable eccentrics (both had been proposed by the Greek astronomers of 300 BC). In the system of deferents and epicycles, the motion of each planet (or the Sun and Moon) is described by a deferent and an epicycle. A deferent is a large circle centred on Earth, and an epicycle is a smaller circle whose centre moves on the deferent. The Sun, Moon and the planets were assumed to move on their epicycle. In the system of movable eccentric there is only one circle per planet (or the Sun and Moon) which moves around a point displaced from the Earth. Individually these two schemes do not fully accounted for all observed planetary phenomena.

Ptolemy combined the two systems. He proposed that each planet rotates on an epicycle whose centre moves on a deferent. But the planet describes uniform circular motion around an equant. The equant is an imaginary point on the diameter of the deferent, located opposite the Earth from the centre of the deferent and the distance between Earth and the centre of the deferent and that between the equant and the centre are equal (Figure 1.1). The uniform motion of the planet is obtained by a suitable choice of the diameter of the epicycle and the deferent and the separation between the equant and the centre of the deferent. This scheme broke with the main assumption of ancient astronomy because it separated the condition of uniform motion from that of constant distance from the centre. Only from the equant will a planet appear to move uniformly; from the Earth and the centre of the deferent the motion will be non-uniform. Because of this departure from the Aristotelian ideal, Arab astronomers, from the eleventh century onwards, challenged the Ptolemaic scheme (see later in this chapter).

In this scheme the plane of the ecliptic is that of the Sun's apparent annual path among the stars. The plane of the deferent of a planet is inclined at a small angle to the plane of the ecliptic. The plane of the epicycle is assumed to be inclined by an equal amount to the plane of

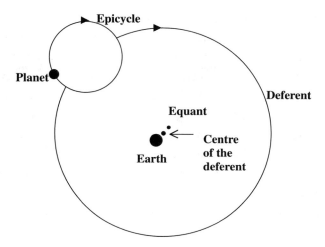

FIGURE 1.1 The Ptolemaic scheme of a planetary orbit. The planets revolve around the Earth but the orbits are not centred on the Earth. The orbit of each planet has two components: there is a circular motion on an *epicycle*, but the centre of the epicycle moves with circular motion on a *deferent*. A planet describes uniform circular motion around an *equant* i.e. the angle planet–equant–earth changes by equal amount in equal time. The equant is an imaginary point on the diameter of the deferent. The Earth and the equant are equidistant from the centre of the deferent.

the deferent, so that the plane of the epicycle of a planet is always parallel to that of the ecliptic. The planes of the deferents of Mercury and Venus are assumed to oscillate above and below the plane of the ecliptic, and the planes of their epicycles similarly oscillate with respect to the planes of their deferents. This complicated scheme gave a better description of planetary motion than had been possible before. But even with this scheme there was considerable difference between the observed and the predicted motion of the Moon. In the fifteenth century this disagreement led Copernicus to doubt the Ptolemaic scheme and to develop his own theory, to be discussed in the next chapter.

Ptolemy believed in the physical reality of the crystalline spheres to which the heavenly bodies were attached. However, he does seem to have realised that the planets were much closer to Earth than the stars.

Beyond the sphere of the stars Ptolemy proposed other spheres ending with the *primum mobile* – the prime mover – which provided the motive power for the universe.

With the decline of Alexandrine scholarship, intellectual activity in the Mediterranean world gradually came to an end. Rome had established its protectorates throughout the eastern Mediterranean by about 200 BC. The Romans were fascinated by the Greeks and were deeply affected by Hellenic philosophy. Most educated Romans learned Greek and yet the scientific idea, the conception of a logical universe, remained alien and exotic to them. In six centuries, despite the Alexandrine example, Rome produced no men of science. It would be wrong to assume that the Romans, and Latins in general, lacked originality and creativity, for their contributions to philosophy and literature were both original and significant. The reason for Rome's failure to produce men of science has to be found in the Roman character and the Stoic philosophy they favoured. The Stoics laid great stress on correct conduct and duty. They believed in a rigid interrelation between different parts of the world, and acquisition of new knowledge was not encouraged. In Rome, therefore, we find a very efficient but unimaginative administration.

Notable in the closing phase of Greek science was the Egyptian Neoplatonist philosopher Hypatia (379–415 AD), the first outstanding woman of science. Daughter of Theon, himself a noted mathematician and philosopher, Hypatia became a recognised head of the Neoplatonic school at Alexandria. Her eloquence and beauty, combined with her remarkable intellectual gifts, attracted a large number of pupils. She wrote commentaries on the *Arithmetica* of Diophantus of Alexandria, on the *Conics* of Apollonius of Perga and on the astronomy of Ptolemy. These works are lost now but their titles have come down to us through the letters of one of Hypatia's pupils. Hypatia symbolised learning and science, which were identified with paganism by early Christians, and she became the focal point of tension and riots between Christians and non-Christians in Alexandria. In 415 AD she was dragged through the streets, and her body was dismembered and burned by rioting

Nitrian monks and their Christian followers. The great library with its priceless collection of books was also burned down. Following the destruction of this centre of learning many scholars left Alexandria, and the Mediterranean world sank into the despair and darkness of dogma, intolerance and persecution. Greek science was lost from view in the first few centuries of the Christian era; superstition and mysticism replaced it. Rational thought gave way to divine revelation as the test for truth, and intellectual activities hardly existed, let alone flourished. Rescue came in the eighth century in the wake of Islam.

ISLAMIC SCIENCE

Some time in the seventh century, after the prophet Mohammed (570–632 AD) had established Islam (622 AD), the Arab armies stormed out of what is modern-day Saudi Arabia. In two hundred years they established an Islamic empire which covered the whole of the Middle East, North Africa and most of Spain and eventually extended to India. There followed a period of four hundred years of peace and stability in which intellectual activity flourished. To the Arabs knowledge was a precious treasure. The Qur'an, the scripture of Islam, particularly praised medicine as an art close to God. Astronomy and astrology were believed to provide the means to glimpse what God willed for mankind. In 762 AD the second Abbasid caliph, al-Mansur, moved his capital from Damascus to Baghdad and began a building programme to transform Baghdad into a new Alexandria. This ambitious programme of construction was continued by the most vigorous patron of arts and sciences, Caliph al-Ma'mun (809–833 AD). His building programme included an observatory, a library and an institute for translation and research named Bait al-Hikma (House of Wisdom). Here Muslim, Christian, Jewish and pagan scholars undertook the monumental task of translating Greek writings from antiquity. In just a few decades major works, including those of Aristotle, Euclid, Ptolemy and Archimedes, were translated from Greek to Arabic. Greek medicine, astronomy, astrology and mathematics, together with the great philosophical works of Plato and particularly Aristotle, were assimilated into Islam by the end of the ninth century. The Arab scholars also borrowed freely from their eastern neighbours in India. From them they learned trigonometric procedures and the decimal number system. But contrary to popular misconceptions, the Arabs did not just passively borrow from other cultures. They criticised and they innovated. Islamic scholars went far beyond the Greek and Indian mathematical methods. Of the six modern trigonometric functions – sine, cosine, tangent, cotangent, secant and cosecant – five are of Arab origin; only the sine function was introduced from India. Arab development in spherical trigonometry was rapid and discoveries in trigonometric identities made solving problems in spherical geometry simpler and quicker. Euclid's parallel postulate came under intense scrutiny. Numbers fascinated Islamic scholars and this served as the motivation to the invention of algebra (from Arabic *al-jabr*) and the study of algebraic functions. Major theses were also produced on optics, which included an early account of refraction and a mathematical approach to finding the focal point of a concave mirror.

Arabs freely applied mathematical methods to astronomy, a subject to which they contributed hugely in theory, computation and instrumentation. The great astronomical observatories they built provided accurate observations against which Ptolemaic predictions could be checked. The first revision of the star catalogue given in Ptolemy's *Almagest* was undertaken in the tenth century by 'Abd ar-Rahman al-Sufin, a Persian astronomer who worked in both Iran and Baghdad. No new stars were added to the catalogue but al-Sufin did note the existence of a faint extended object in the constellation of Andromeda. This is the earliest known record of a nebula. Today we know this 'patch' as the great Andromeda spiral galaxy, and in the twentieth century it was to play a crucial part in establishing the scale of the universe. In the following centuries, Arab astronomers made observations of new stars and reobserved the planets to improve the values of astronomical parameters. However, the Arabs worked strictly within the Aristotelian and Ptolemaic framework of a geocentric universe. Arab astronomers actually went to considerable lengths to remain faithful to Aristotle and great efforts were made to follow the Ptolemaic approach. New astronomical observations were made, but they were only used to improve the values of the parameters of planetary orbits as defined by Ptolemy. In the tenth century Mahammad al-Battani (929 AD) repeated Ptolemy's observations in a searching and exacting manner. From this he constructed his own astronomical table. This was the greatest astronomical work since the *Almagest*, which it ultimately replaced. A Latin translation reached medieval Europe where it was printed and distributed widely. In the fifteenth century Copernicus used it extensively in his great work *On the Revolution of the Heavenly Bodies* (see Chapter 2, Kepler).

Criticism of Ptolemaic cosmology emerged in the eleventh century. Ibn al-Haytham (965–1039 AD), a leading philosopher in Cairo argued, in his book called *Doubts on Ptolemy*, that the equant (see Figure 1.1) failed to satisfy the requirement of uniform circular motion. He attempted to discover the physical reality behind Ptolemy's mathematical model and conceived of a single celestial sphere for each component of Ptolemy's planetary motions. This work reached Europe in the fourteenth century and had major influence in the early Renaissance Europe. Even more severe criticism of Ptolemy's cosmology was to follow. In the twelfth century, Ibn Rushd (1126–1198 AD), born in Cordoba, Spain, and known in Europe as Averroes, rejected Ptolemy's eccentric deferents and argued for a strictly concentric model of the universe. However, attempts to formulate such a model failed. Further refinements were made in the thirteenth century in Iran where Nasir al-Din al-Tusi (1201–1274 AD) added two more epicycles to each planet's orbit. He succeeded in reproducing the nonuniform motion of planets by a combination of uniform circular motions. Later this scheme was extended to obtain a perfectly geocentric scheme of planetary motion.

The philosophical objections to Ptolemy's model of planetary motion identified by the Islamic scholars were similar to those of Copernicus two hundred years later. But the Islamic scholars did not take the next logical step and propose a heliocentric model of the solar system. The model proposed by Copernicus would have appalled them, but they would have understood the need for such a model. Islamic scholars nevertheless opened the door to question Ptolemy and his world-view.

Islamic Spain was a storehouse of knowledge for European scholars. In Cordoba, the capital of Andalusia, there were bookshops and libraries, then unknown in Christian Europe. The central library built in the *Alcazar* (the Royal Palace) around 970 AD had over 400,000 titles. This rich and sophisticated society took a tolerant view of other faiths and encouraged scholarship. Jewish and Christian scholars lived and worked in peace and harmony with their Islamic counterparts. Through translations of Arabic works, the Indian decimal system of numeration was transmitted to Europe. Ptolemy's *Almagest* and various works of Aristotle and other Greek authors were translated from Arabic to Latin. These translations and also Greek works translated directly into Latin rekindled the spark of inquiry in Europe. Today it appears ironic to read Ibn Khaldun who wrote in 1377 AD that 'we have heard of late that in Western Europe the philosophical sciences are thriving, their works reviving, their sessions of study increasing, with abundant teachers and students'. However, the world-view transmitted by Arabs to the Christian Europe may have been new in tone but was old in substance. The Latin Christians were Neoplatonic in their thought, with an Aristotelian view of both the micro- and macrocosmos. The study of Arabic works sharpened this view and the Aristotelian view of the structure of the universe dominated the Christian world-view from the thirteenth century onwards. During this period astrology became a central intellectual interest and it retained this prominent position until the triumph of the experimental method in the seventeenth century.

The Greeks have handed down to us two methodologies for investigating nature. The 'Platonic' tradition emphasises the unchanging nature of the fundamental laws and regards observable events as a consequence of these laws. The 'Aristotelian' view, on the other hand, emphasises the observable nature with its complexities rather than the unobservable 'laws' which are responsible for these complexities. The Church Fathers of the eighteenth and nineteenth century interpreted the 'Platonic' laws as proof of the existence of God. These Platonic and Aristotelian methods have proved remarkably durable. They have survived over a period of more than two thousand years and form the backbone of current scientific thought.

2   Kepler

The Arab instrumentation, observations, astronomical tables and maps retained their superiority at least until the middle of the thirteenth century. But the diffusion of Greek literature from the Middle East and Spain gradually rekindled the spirit of inquiry in Europe. In the twelfth and the thirteenth centuries the writings of Aristotle on physics, metaphysics and ethics became available in Latin, translated from either Greek or Arab sources. These were crucial for the greatest of the medieval Christian thinkers, St Thomas Aquinas (1225–1274 AD). Aquinas studied Aristotle in great detail and wrote numerous commentaries on a variety of Aristotle's works. One of the Aristotelian themes that influenced Aquinas was that knowledge is not innate but is gained from the senses and from logical inference of self-evident truths. To this Aquinas added divine revelation as an additional basis for inference. Aquinas used Aristotle's dictum that everything is moved by something else to argue that the observable order of cause and effect is not self-explanatory. It can only be explained by existence of the 'First Cause' or God. The concept of the 'First Cause' can be traced back to the Greek thinkers and had become an underlying assumption in the Judeo-Christian world-view. The argument for the existence of God inferred from motion was given doctrinal status in the first two 'proofs of God' of Aquinas:

- Things are in motion, hence there is a first mover
- Things are caused, hence there is a first cause

But where Aristotle was concerned with understanding how the world functions, Aquinas was concerned more fundamentally with explaining why it exists. There was, however, no conscious conflict between science and religion in the Middle Ages. As Aquinas noted, God was the author of both the book of Scripture and the book of nature. But by weaving knowledge of nature into his theology, as in the proof of the existence of God from motion, Aquinas had laid the foundation for a possible conflict between science and religion. A challenge to his scientific 'proof' of motion would necessarily be perceived as a theological challenge.

Before the twelfth century the European view of life and the universe was unquestioning and mystical. St Augustine summarised this in his *'Credo ut intelligam'* (understanding through belief). The new Arab knowledge changed all this: *'Intelligo ut credam'* (belief through knowledge) became increasingly acceptable. One of the first scholars to visit Spain and return with his discoveries was an Englishman from Bath called Adelard, who had travelled in the Muslim countries of the Middle East before arriving in Spain in the second decade of the twelfth century. His main interest was astronomy and when he returned to England he had with him a Latin version of an Arab translation of Euclid's *Elements*. But more importantly he carried with him the new method of thought he found in the Arab academic world. Adelard had acquired rationalism and the secular, investigative approach typical of Arab natural science. Adelard's insight convinced him of the power of reasoning, rather than the blind respect for all past authority that he had encountered in Latin Europe.

Following Adelard, other European scholars went to Spain and other Muslim countries and returned with manuscripts and new methods of thought. In this new thinking the central figure was not Christ and 'the man' was no longer the transient, worthless figure described by the medieval theologians, but an independent, intelligent and capable, forward-looking individual. This concentration on the human rather than the divine became known as humanism and this secular doctrine spread throughout Europe in the fourteenth and fifteenth century. The humanists, as a class, were not sympathetic to the scientific outlook but their questioning attitude and the contacts that developed between craftsmen and scholars in the early Renaissance were to change the nature of scientific inquiry in Europe. As the questioning grew, knowledge became institutionalised and this crystallised in the establishment of European universities. In these universities the students were taught to think 'investigatively' rather than to follow dogma.

Georg Peuerbach (1423–1461 AD) was born in Vienna and studied in Italy. He devoted himself to the study of Ptolemy's *Almagest*; he had acquired a copy of an Arab translation. Because of the profusion of errors in the Latin translation, Peuerbach began a new translation, from the original Greek. Peuerbach unfortunately died at the early age of 38 but he had a very able pupil, Johannes Müller (1436–1476 AD), better known as Regiomontanus after the Latinised form of Königsberg, his birthplace. After Peuerbach's death, Regiomontanus continued his work and in 1461 AD travelled to Rome to learn Greek. He also collected manuscripts from the Greeks who had fled from Constantinople after it fell to the Turks in 1453 AD. Among these manuscripts was a Greek text of the *Almagest* and with this Regiomontanus was able to complete the work started by Peuerbach. In 1472 AD he moved to Nürnberg, where he made his first scientific observations, his subject being a comet. Regiomontanus was primarily a mathematician and in 1464 AD he wrote *De triangulis omnimodis libri quinque* (Five books on triangles of all kinds). This was the first entirely mathematical book (that is, it was independent of astronomy) written in Europe on plane and spherical trigonometry. The work used methods of presentation applicable to general classes of problems on triangles and used algebraic techniques to simplify their solutions. This book was published 57 years after Regiomontanus's death. In 1474 AD Regiomontanus had published his *Ephemerides*, navigational tables showing daily positions of the celestial bodies for several years, which Columbus apparently used on his voyage of European discovery of the Americas. In 1475 AD Regiomontanus was summoned to Rome by Pope Sixtus IV for the long-contemplated revision of the calendar. Without adequate observational data on planets and stars there was little he could do and the reform of the calendar was deferred for another century. Regiomontanus died in Rome, possibly from the plague. He was chiefly responsible for the revival and advancement of trigonometry in Europe and with Peuerbach he had created the tools for Copernicus to 'change the universe'.

THE COPERNICAN REVOLUTION

The Polish cleric-astronomer Nicolaus Copernicus (1474–1543 AD) was born in the town of Torun on the banks of the river Vistula. His father, who was a merchant, died when Nicolaus was about ten years old and he was brought up by a benevolent uncle who later became a bishop. Nicolaus's early education was at the school in Torun and the renowned Polish University of Cracow where his studies included mathematics and astronomy. His uncle wanted him to enter the canonry of Frauenburg in order to secure lifelong financial security. While waiting for a vacancy at Frauenburg he was sent to Italy to broaden his education. Here, at 23 years old, he entered the University of Bologna, a renowned centre of learning, where his studies included classics, law, mathematics, astronomy, medicine and theology. He is also said to have dabbled in art and painting, suggesting a skill in visualisation often associated with scientific minds. In Bologna he made his first astronomical observation, an occultation of the star Aldebaran by the Moon, on 9 March 1497. The same year he was also elected (by proxy) a canon of Frauenburg, but he decided to stay in Italy to continue his studies. He enrolled at the University of Padua to study both law and medicine.

In 1503 AD the University of Ferrara awarded Copernicus the degree of doctor of canon law and that year he returned to Poland as a canon of the cathedral of Frauenburg, in which capacity he served the Church for the rest of his life. For Copernicus astronomy was an intellectual hobby. He had translations of the *Almagest* and the tables of al-Battani, Peuerbach and Regiomontanus and he decided to study the then known theories of planetary motion. Between 1497 AD and 1529 AD he made enough observations to enable him to recalculate the major components of the supposed orbits of the Sun, Moon and planets around the Earth. He found the complexity of the theories of orbits, their internal contradictions, their disagreements with each other and the disagreement with the available data unacceptable. This work rapidly enhanced his reputation as an astronomer and in 1514 AD he was invited by Pope Leo X at the Lateran Council to give his opinion on the proposed reform of the calendar. This he refused to do as the lengths of the year and the month and the motion of the Sun and Moon were not known with sufficient accuracy to permit a proper reassessment.

As his astronomical studies progressed he became increasingly dissatisfied with the Ptolemaic system of planetary motion. To fit the observations of planetary motions, Ptolemy had been forced to offset the centres of regular motion of the planets slightly away from Earth (see Chapter 1, Aristotle). This Copernicus believed was in conflict with the basic rule of true circular motion required by the Aristotelian doctrine. This had also been the argument used by Arab scholars in the twelfth century. He therefore turned to the scheme proposed by the Greek philosophers of antiquity, Philolaus (who died in 390 BC), and Aristarchus (310–230 BC). He was to write in the introduction of his *De Revolutionibus*[1]:

> Hence I thought that I too would be readily permitted to ascertain whether explanations sounder than those of my predecessors would be found for the revolution of the celestial spheres on the assumption of some motion of the earth.

Copernicus revived the model of the solar system proposed by Aristarchus with the Sun at the centre. He released the Earth from its (Aristotelian) static, central position and set it in motion, both around the Sun and on its own axis. The idea of a moving Earth seemed absurd and dangerous and flew in the face of common sense. But his scheme immediately explained the greatest problem faced by the geocentric system – that of the apparent back-and-forth movement (retrograde motion) of planets. He was also able to explain the variation in planetary brightness by a combination of motion of the Earth and the planet. Copernicus explained why Mercury and Venus are never seen opposite the Sun in the sky: he placed their orbits closer to the Sun than that of the Earth. He placed the planets in order of their distance from the Sun by considering their speeds. But the theory still assumed that the universe was spherical (because 'the sphere is the most perfect') and finite, terminating in the sphere of the fixed stars. Because he insisted that planetary motion is circular at uniform speed, he had to retain the epicycles and the eccentric deferents but now centred close to the Sun. He also retained the crystalline spheres. Thus his model was almost as complicated as that of Ptolemy, requiring 34 circles to explain the motion of the celestial bodies. But Copernicus found it aesthetically satisfying, and aesthetic considerations are not to be ignored in science. He also believed that it was the true picture of the divinely ordained cosmos. However, there were implications of the theory that caused considerable concern. Why should the crystalline orb containing the Earth circle the Sun? Why were objects including humans not hurled off the surface of a rotating Earth? Copernicus was a high and respected official of the Church and he must have been aware of the dangers of differing from the Christian orthodoxy of the time. He must also have known of the challenge to the established Church being mounted by Martin Luther in neighbouring Germany and of the charges of impiety levelled against Philolaus and Aristarchus. Copernicus wisely decided not to publish his theory of the Sun-centred universe. In May 1514 he prepared a short paper and circulated it privately among his friends. The paper bore the title *De hypothesibus motuum coelestium a se constitutis commentariolus* (A Commentary on the Theories of the Motion of Heavenly Objects from their Arrangements). The paper summarised his new idea that the apparent daily motion of the stars, the annual motion of the Sun, and the retrograde motion of planets result from Earth's daily rotation on its axis[2] and its annual revolution around the Sun. The Sun was assumed to be stationary at the centre of the planetary system. The Earth, therefore, is the centre, not of the universe, but only of the Moon's orbit. In the subsequent years he developed his arguments with mathematical computations and diagrams. Lectures on the ideas proposed in the *Commentary* were given in Rome in 1533 AD before Pope Clement VII, who approved his work, and a formal request was made, in 1536 AD, to have the *Commentary* published. But Copernicus continued to hesitate because of 'the scorn which I had reason to fear on account of the novelty and unconventionality of my opinion.'[1]

Finally in 1540 AD, urged on and encouraged by Nicholas Schönberg, the cardinal of Capua, and Tiedemann Giese, the bishop of Chelmno, Copernicus gave the completed manuscript of the book now entitled *De Revolutionibus Orbium Coelestium* (*The Revolutions of The Heavenly Orbs*) to Georg Joachim Rhäticus, his pupil, to take to Nürnberg, Germany, for publication. But Rhäticus passed the task of publication to Andreas Osiander, a leading Protestant theologian and one of the early fathers of the Lutheran creed. In *De Revolutionibus* Copernicus reiterated his proposal that the centre of the universe was not the Earth but a spot somewhere close to the Sun. He also rejected the arguments for a stationary Earth and the old arrangement of the planets – Earth, Moon, Mercury, Venus, Sun, Mars, Jupiter and Saturn. Instead he proposed a heliocentric system with a stationary Sun and planets arranged in the following order: Mercury, Venus, Earth with the Moon orbiting round it, Mars, Jupiter and Saturn (Figure 2.1). His model of the solar system made possible accurate predictions of lunar motion for use in the proposed reform of the calendar and, as we have seen, explained the planets' variation of brightness, retrograde motion and velocity. Copernicus explained the tricky problem of the absence of stellar parallax (that is, any apparent periodic shift in the position of fixed stars as the Earth moves in its orbit around the Sun) inherent in a Sun-centred universe, by stating that the stars were at such enormous distances that their parallax was immeasurably small. Copernicus thus accepted a vast cosmos consisting mostly of empty space. But he side-stepped the question of the motion on Earth and why an object thrown into the air from a revolving Earth did not fall to ground to the

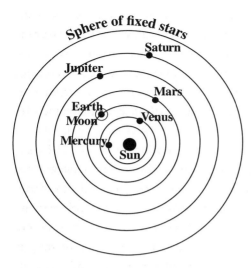

FIGURE 2.1 The Copernican universe. The planets revolve around the Sun or more accurately around a point offset from the Sun, similar to the equant in the Ptolemaic system. The Moon revolves around the Earth. In this scheme the orbits are perfect circles and planets move with uniform velocity. The simplicity of this diagram is rather misleading because to reproduce the observed motion of the planets a number of epicycles were incorporated and the real scheme is almost as complicated as that proposed by Ptolemy.

west. A copy of *De Revolutionibus* is believed to have been brought to Copernicus at Frauenburg on the last day of his life, 24 May 1543.

Osiander wrote the preface of *De Revolutionibus* and inserted it anonymously; it has often been attributed to Copernicus. In it Osiander states[1]:

> For these hypotheses need not be true or even probable. On the contrary, if they provide a calculus consistent with the observations, that alone is enough.

That is, the theory should be treated as a hypothesis and a mathematical exercise and should not be taken to represent physical reality. He then went on to imply that some of the consequences of the theory were absurd. One has to suspect that Copernicus had not seen or approved the preface.

The Copernican model appealed to a large number of independent-minded astronomers and mathematicians, both for its elegance and because it broke with the Aristotelian tradition, which they felt hampered development. The Aristotelian tradition and the Ptolemaic geocentric scheme were at the heart of Western Christian thought and both views had been elevated almost to the level of religious dogma.

Surprisingly, the Copernican scheme was acceptable to the Roman Church. The Church, sitting in council at Trento, accepted the text of *De Revolutionibus* without reaction. It was unconcerned with the revolutionary nature of the Copernican thought and his scheme was used, in 1582, to reform the calendar. The first attack came from the Protestants. Martin Luther said: '. . . the fool wishes to reverse the entire science of astronomy'.

Copernicus' heliocentric universe struck at the heart of Christian belief. He removed the Earth from the centre of the universe and so from the focus of God's purpose. In doing so, Copernicus destroyed the doctrine of 'natural motion' and 'natural place'. The Aristotelian view was that bodies fell to their 'natural place', which was the centre of the universe – the Earth. But in a heliocentric theory, the Earth no longer coincided with this centre and the fall of bodies to the ground made no sense. Although Copernicus swept aside Aristotelian mechanics he did not suggest anything to put in its place. Thus, for those who wanted to promote Copernicus's ideas, the question of why the motion of Earth was not noticed took on a special urgency. The re-examination of the laws of falling bodies led eventually to the concept of universal gravitation. Copernicus also changed the size of the universe. He claimed that the lack of stellar parallax suggested that the starry sphere was at a vast distance; indeed, he came close to saying that the universe was infinite. This scheme implied an enormous cosmos consisting mostly of empty space. The Christian doctrine asserted that God did nothing in vain, so what was the purpose of immense empty space in which the Earth and mankind had been set in motion?

The teachings of Copernicus had little influence on contemporary scientific thought. His theory was mentioned rather infrequently but not always unsympathetically. The Copernican system was seen as a convenient mathematical fiction rather than physical reality. Although the Christian Church had adopted the Aristotelian scheme of the heavens, by the Middle Ages it had been able to accommodate a spherical rotating Earth. It is possible that eventually the Church could have come to terms with the Copernican scheme, since a world system with the Sun at the centre was no less in accord with biblical thought than a spherical Earth. But the fate of the Copernican system was, unfortunately, determined by events taking place outside the scientific world. The Christian world was breaking apart into the Catholic and Protestant factions and it was doing so in a profoundly acrimonious manner. Into this turbulent world stepped the tragic figure of Giordano Bruno.

Giordano Bruno (1547–1600 AD) was born near Naples in Italy and was a renegade monk. He was an aloof and lonely man, who taught an obscure and barren system of logic at various universities in Italy and France, and was unwelcome at all of them. In 1583 AD he went to London where he published (but with a false impress of Venice) three small books, *The Ash Wednesday Supper*, *On Cause, Principle and Unity* and *On Infinite Universe and its Worlds*. These books effectively contain his entire world-view, largely based on the works of Copernicus, which were not yet prohibited. Bruno maintained not only that the Earth moves round the Sun but that the Sun itself moves and there is no such thing as a point at absolute rest. The stars are at vast and various distances from the Earth and are themselves centres of other planetary systems. The universe is infinite and can provide no fixed point of rest and our planetary system is in no sense the centre of this universe. Bruno was not a scientist, he had no formal training in sciences and was not familiar with the methodology of science, and yet what he had to say was incomparably more revolutionary than anything written by the conservative Copernicus. It appalled the Christian Church. In medieval Christian philosophy a 'created Universe' was fundamental and the Creator was separate from his Creation. This philosophy had no room for an infinite universe endless in time and space. Bruno was condemned by the Inquisition and spent seven years in various prisons. In 1600 AD he was burned at the stake. Bruno's views, though philosophical and speculative, mark the real change from medieval thought and heralded the approach of modern scientific thinking. Four hundred years later, in February 2000 the Roman Catholic Church issued a statement 'regretting' the burning of Bruno.

THE PROFESSIONAL ASTRONOMER

The heliocentric model of Copernicus was based on the data of Hipparchus, Ptolemy and al-Battani, and improvements to these data by various Arab astronomers. These were the best data available to him. In the next 50 years, the Danish astronomer Tycho Brahe (1546–1601 AD) was to change this. Tycho was born into a wealthy noble family. He was adopted by a rich childless uncle who sent him at the age of 13 to study law at the University of Copenhagen. In Copenhagen he witnessed a partial eclipse of the Sun and this fired his fascination for astronomy. At 17 Tycho was sent to Leipzig to study law but his love for astronomy was so deep that he neglected jurisprudence. Here he started on the career of observational astronomy that was to be his life's work. He bought an astrolabe and made his first observations, one of which was of a conjunction of Jupiter and Saturn. When he compared his observations with the existing tables of planetary and stellar positions he found that there were a number of inaccuracies in the tables. The available astronomical tables were several days off in predicting the conjunction. This realisation convinced Tycho that progress in astronomy would not be possible without highly accurate observations. At 26, after studying at various German universities and Basle in Switzerland, he returned to Denmark. By then he was also a wealthy man as his uncle had died, leaving him a fortune.

With this financial security, Tycho was able to pursue his interest in astronomy by building himself a small observatory. From here, on 11 November 1572, he made careful observations of a 'new star', brighter than Venus, which had appeared in the constellation of Cassiopeia. Tycho measured the position of this star regularly for 18 months till it became too faint for observations by the unaided human eye. He showed that the position of the star did not change relative to the fixed stars, suggesting that it lay in the realm of the stars, but he also noted that the colour of the new star changed from yellow to red as it faded. This new star, which was observed by a number of other astronomers and also by lay persons, shook Aristotelian cosmology to its very foundations. The sixteenth-century European world protected itself against the uncertainties of the future by its confidence in the Aristotelian doctrine of fundamental and continuous harmony in the universe. This harmony was ruled by stars, which were regarded as perfect and unchanging. Changes were supposed to occur only in the local sub-lunar sphere. The appearance and the dramatic change in the new star, together with the reports of the Copernican theory that the Sun and not the Earth was at the centre of the universe, shook confidence in the immutable laws of antiquity.

Tycho published the observations of the new star (now known to have been a supernova in our galaxy) in 1573 AD, in *De nova stella*. At first he was uneasy about publishing his results, as in the sixteenth century it was considered unseemly for a nobleman to write books, and Tycho had no doubts about his noble descent! But the publication of the book transformed Tycho, at 28, from a Danish dilettante to an astronomer with a European reputation. Tycho ends his account of the new star with its astrological implications. He forecast that the influence of the new star would begin to be felt from 1592 and would last until 1632 and that there would be religious upheaval. Many people were impressed by this prediction as the great Protestant champion Gustaf Adolf was born in 1594 and died in 1632. Tycho's book also had immediate and very beneficial consequences: it brought his astronomical talents to the attention of the Danish monarch, Frederick II. The king offered him the island of Hven, near Copenhagen in the sound between Denmark and Sweden, and funds to build and operate a major astronomical observatory. This is probably the most magnificent gift a king has ever made to an astronomer. In 1576 AD, Tycho set about this task in his characteristically extravagant but meticulous style. He was an artist as well as a scientist: everything he undertook or surrounded himself with had to be innovative and beautiful. He imported the best craftsmen, to construct the finest instruments, and artists and architects to design and decorate his observatory. In fact, Tycho built two observatories, one called Uraniborg (in honour of Urania, the Greek goddess of astronomy) and the other Stjerneborg (the Star Castle). Both were equipped with a full complement of instruments and support staff. Tycho and his assistants made observations which substantially corrected nearly every known astronomical record. The observatory also became a meeting place for visiting scholars and learned travellers from all over Europe.

Tycho started a programme of accurate measurement of positions of fixed stars, the Moon and the five planets as they moved over the celestial sphere. The measurements were made independently by his assistants in the two observatories and only he was allowed to look at the two sets of results. This was to remove the bias in measurements introduced by observers. In his analysis Tycho allowed for systematic errors introduced by the flexure of the instruments and by atmospheric refraction when stars are observed at different angles above the horizon. These precise observations were made without the aid of a telescope (Galileo first used a telescope for astronomical observations in 1609 AD, about eight years after Tycho's death). Tycho's instruments consisted of ingenious improvements on ancient devices like the quadrant, armillary sphere and parallactic ruler (some of which are shown in Figure 2.2). Tycho's goal was to establish exact positions of as many fixed stars as possible and to measure accurately the positions of the visible planets. These measurements were made over a period of 20 years and the result was a star catalogue in which the positions of 777 stars (about three-quarters of all stars visible with the naked eye from the latitude of Denmark) were measured with an accuracy of about 1–2 arcminutes. These were the most accurate and consistent measurements made since antiquity; and they were almost 60 times more accurate than the measurements of Hipparchus, Ptolemy and al-Battani. More importantly, the measurements were made consistently and repeated over a long period and the errors in his observations were quoted.

Tycho had studied the works of Copernicus and he did not like the heliocentric model. Tycho had correctly surmised that if the heliocentric model were true then the stars would appear to wobble as they were observed from different positions in the Earth's orbit. This stellar parallax had never been observed and was beyond even Tycho's very accurate measurements. Tycho had a very firm grasp of the complexities and errors involved in observational astronomy and he deduced that if he could not observe the stellar parallax then the stars must be over ten thousand times further away than the Sun. Such distances were beyond his comprehension and this coloured his view of the Copernican model. Tycho was also uneasy with the Ptolemaic model of the solar system with its epicycles centred on circles and not on celestial bodies. He therefore proposed a model with a stationary Earth around which the Sun and the Moon revolved in circular orbits and with constant speed. But the planets revolved round the Sun, also in circular orbits and with constant speed (Figure 2.3). He did away with the sphere of the fixed stars. Although Tycho's model of the solar system differed conceptually from that of Copernicus the two models are mathematically equivalent. This Earth-at-rest scheme was published in *Astronomiae Instauratae Progymnasmata*, perhaps the most important publication of Tycho. Three volumes were planned but only two were published.

In the second volume of *Astronomiae* Tycho published his observations of a comet that appeared in 1577 AD and was visible for 10 weeks.

FIGURE 2.2 Some of the instruments used by Tycho Brahe for his astronomical observations. The full description of all the instruments and how they were used is given in Brahe's *Astronomiae Instauratae Mechanica*. The four instruments in this figure are: (a) Double arc instrument for measuring angular distances. This instrument was used to measure distances of up to one-twelfth of the circumference of the circle, or one sign of the zodiac. (b) Great equatorial armillary instrument with one complete circle and one semi-circle. This instrument was used to measure the declination and relative right ascension of a star. The semi-circle could also be used to measure the equatorial distance from the meridian, and from this the time could be determined from the known positions of the star and the Sun. (c) Azimuth quadrant, used to determine the altitude as well as azimuth of a star. A star was viewed through the tube DE; the altitude was measured along a vertical scale (shaped as a quarter section of a circle) and the azimuth along the horizontal scale. (d) Equatorial armillary sphere with movable equator ring. The armillary sphere is the oldest known astronomical instrument and consisted essentially of a skeletal celestial globe whose rings represent the great circles of the heavens. Tycho Brahe used it to determine the declinations and the right ascensions of the stars. (Pictures provided by National Museum of Science & Industry/Science & Society Picture Library.)

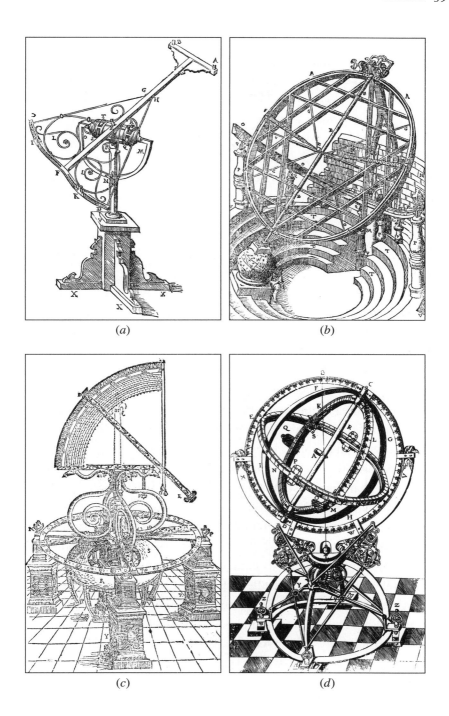

(a) (b) (c) (d)

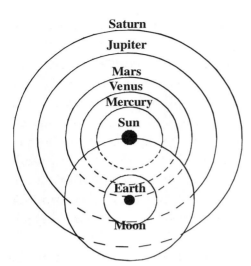

FIGURE 2.3 Tycho Brahe's solar system. Both the Moon and the Sun revolve around the Earth on circular orbits but the other planets revolve around the Sun, also on circular orbits. Mathematically this scheme is equivalent to that proposed by Copernicus.

Tycho made accurate measurements of the positions of this comet, and he combined these with measurements made by other astronomers in Europe (for astrological purposes) to show that the parallax of the comet was very small. This led him to conclude that the comet was more distant than the Moon. Comets had hitherto been considered to be part of the earthly, sub-lunar world. Tycho's observations conclusively demolished this notion. This then was another change in the changeless heavens and another blow to Aristotelian cosmology. Tycho also showed that the comet was moving on an elliptical path and not a circle, the path that the celestial bodies were expected (almost required) to follow. Tycho published his findings in 1588 and he concluded 'There are not really any spheres in the Heavens'. Tycho left unanswered the really difficult question of what kept the planets and comets in orbit and why the noncircular orbit of a comet does not become unstable.

In 1598 AD King Frederick II, the patron and friend of Tycho, died and was succeeded by his son, Christian. The drain on the royal exchequer imposed by Tycho's astronomical enterprise alarmed Christian. He was also unhappy with Tycho's vain and arrogant attitude and his rather extravagant life-style. An impertinent letter from Tycho, demanding additional funding and threatening to leave Denmark if the support was not forthcoming, compounded this situation. Christian conceded nothing and stated plainly that the future funding and patronage would depend entirely on a major change in Tycho's behaviour and attitude. Tycho realised that his blunder had lost him the royal support and that he would have to find another sponsor if he was to continue his observational programme. He found one in 1599 AD and moved to Prague, to the court of Rudolph II, as an Imperial Mathematician. Rudolph offered Tycho the castle of Benatek, northeast of Prague, together with funding to build another great observatory and to transport and install the instrumentation from Denmark. Tycho attempted to continue his observational programme at Prague but the spark of inquiry was no longer there.

During his final years in Denmark, Tycho received a letter from a young mathematician teaching at an obscure school in Graz in Austria. The letter proposed an orderly and mathematical scheme of the solar system based on the Copernican heliocentric model. Although Tycho did not believe in the Copernican model, he was sufficiently impressed by the young German to encourage him, instead, to work on the model of the solar system which he himself had proposed. He also invited him to join him at Prague. The young mathematician's name was Johannes Kepler. Kepler, it seems, was glad to leave Graz and join Tycho in Prague. But he had a difficult relationship with Tycho. He found that he did not have the full and easy access to the data that he desired. Tycho zealously guarded his data and would only release them in drips and drabs. In 1601 AD Tycho died after a bout of prodigious eating and drinking at a banquet given by a local baron. Even in death Tycho revealed his flamboyant nature by willing his astronomical data to Kepler, data which he had parsimoniously withheld from Kepler during his lifetime.

Tycho Brahe was an exceptionally lucky astronomer. Within his lifetime he was able to observe both a galactic supernova and a bright comet. There are not many astronomers who have such luck. He obtained the most accurate and consistent astronomical data on the motion of planets. These data, obtained with simple instruments (by modern standards) but with exceptional practical talent, were an outstanding accomplishment of the Renaissance. But he failed to find the 'secrets of the universe' which he craved. The major questions thrown up by his observations were: if the planets did not move on crystalline spheres, why did they not fall, and what was the medium in which they moved? These questions remained unanswered. Nevertheless his achievements were remarkable; his star catalogue was edited and published posthumously by Kepler who added another 286 stars to the list. This catalogue is known as the Rudolphine Tables after his Polish patron Emperor Rudolf II. But of fundamental importance were his measurements of the perturbations in the motion of the Moon and the measurements of motions of planets. These would profoundly alter our understanding of the solar system and set the stage for the formulation of the theory of universal gravitation.

Tycho made the last great attempt to represent the universe according to the ideal form of a circle, in the Pythagorean spirit. Kepler started with the Pythagorean notion but then abandoned it, and there was a firm and permanent break with the Greek tradition. Before Tycho the distinction between stars and planets was purely phenomenological. Tycho removed the divine sphere and the sphere of stars from his scheme and identified the planets, the Sun, the Moon and the Earth as a single structure, independent of the stars.

MARRIAGE OF MATHEMATICS AND OBSERVATIONS

Johannes Kepler (1571–1630 AD) changed the ancient geometrical description of the heavens to the modern dynamical astronomy, into which he also introduced the concept of a physical force. His worldview was essentially Platonic; he was convinced that the complex motions of the planets must correspond to underlying simple laws, which must be expressible in numeric or geometric form. Born into a poor family in a small town near Württemburg in south-west Germany, Kepler would have received no education at all but for the enlightened dukes of Württemburg who had introduced a system of grants and scholarships to educate gifted children from impoverished homes. Kepler won the necessary scholarships, which enabled him to study at schools and seminaries and eventually at the University of Tübingen. At Tübingen he was taught astronomy by a teacher who, unusually for his day, believed that the Copernican system of the universe was basically true. Kepler enthusiastically adopted the Copernican doctrine and mounted a spirited defence of it in a public debate – not a wise move for a would-be Lutheran pastor! In 1594, during his last year of training in theology, the university strongly recommended him for a position (to teach mathematics) in the Lutheran high school at Graz, in Austria. Kepler left Tübingen to take up this teaching post, without finishing his theology course.

At 20 Kepler began teaching mathematics in Graz. He spent his first year studying the subject that he was supposed to teach. While studying and teaching geometry Kepler had an inspiration; there are only five intervals between the six (then) known planets, and there are also only five regular solid figures, that is figures with equal sides and angles – the Platonic bodies. These regular solids can each be perfectly circumscribed by a sphere (that is, a sphere that would touch each apex of the solid) and a sphere can also fit perfectly inside each of the solids (touching each side of the solid). Kepler, a firm adherent of the Pythagorean world-view and sustained by visions of mathematical harmonies in the skies, proposed that the five regular solids could be fitted between the spheres of the six known planets. His nest of alternating planets and regular solids, described in his *Prodromus Mathematicarum Continens Mysterium Cosmographicum* (Cosmographic Mysteries), was published in 1596 AD. His scheme of planets was as follows[3]:

> The Earth's orbit is a measure of all things; circumscribe around it a dodecahedron [12-sided figure] and the circle containing it will be Mars; circumscribe around Mars a tetrahedron [4-sided figure], and the circle containing it will be Jupiter; circumscribe around Jupiter a cube, and the circle containing it will be Saturn. Now inscribe within the Earth an icosahedron [20-sided figure], and the circle containing it will be Venus; inscribe within Venus an octahedron [8-sided figure], and the circle containing it will be Mercury. You now have the reason for the number of the planets.

Kepler had thus introduced, for the first time, a unitary theory to explain the structure of the universe. Remarkably, it could account for the relative radial distances of planets from the Sun within about 5%. Kepler sent copies of this, his first major work on astronomy, to a number of scientists in Europe including Tycho Brahe. By 1619 AD Kepler had grown uneasy with the 5% discrepancy in the planetary distances and he abandoned his geometric model. But he retained his belief in the Pythagorean single principle or a fundamental law to explain the structure of the universe. This was based, like that of most medieval thinkers, on the belief that the physical universe had a moral plan, a divine purpose, and any mathematical relationship was an illustration of this plan.

In 1599, as we have seen, Kepler was invited by Tycho Brahe to join him as an assistant in his new observatory near Prague, an invitation which Kepler readily accepted. Kepler's decision to leave Graz was made both to have access to Tycho's data, which Kepler needed to continue work on his model of the universe, and also to escape from the 'theological cleansing' of the Lutherans pursued by the Catholic Archduke Ferdinand of Austria. However, Tycho and Kepler had what can at best be described as a difficult working relationship. Although their astronomical skills were complementary, in every other way they could not have been more different. Tycho was a rich aristocrat, Kepler was from a poor family. Tycho was a well known astronomer, Kepler was a beginner. Tycho was a hard-eating, hard-drinking gregarious man, Kepler was a shy introvert whose eating habits can best be described as frugal. Worse, Kepler had expected free and easy access to Tycho's data (which had not been published) but soon realised that Tycho regarded these data as his personal treasure and would only release them in small quantities when he felt exceptionally generous. Moreover, Kepler wanted 'academic freedom', to pick and choose astronomical problems to work on and compare his computations against Tycho's data. Tycho wanted him to work on problems that he, Tycho, considered interesting and important. For example, an outstanding problem of the time, and one that had defied Tycho, was that of the motion of the planet Mars. Kepler volunteered to work on this and Tycho seems to have reluctantly agreed, but the data he released were inadequate for the task. Many acrimonious disputes ensued. To add to this, Kepler was concerned about his salary, which was not sufficient for him to bring his wife from Graz to Prague. After five months, Kepler had had enough and he decided to return to Graz, unsure of his future. But the conditions in Graz had worsened; the Archduke had stepped up religious persecution. Kepler was called before an ecclesiastical court and was invited to renounce his Lutheran faith for Roman Catholicism. Kepler declined and in desperation wrote to Tycho explaining his predicament. Tycho generously invited Kepler back to Prague as his senior assistant, a permanent position approved by the Emperor, but could not resist a display of his authority: Kepler was to be his assistant in the true sense of the word, and was to work only on the problems assigned to him by Tycho. Kepler had no choice but to agree. He returned to Prague with his wife and Austria lost, to religious bigotry, one of the greatest astronomers in history.

In Prague Kepler settled into a stable working relationship with Tycho, working on problems which he found boring and which kept him away from his passionate interests: the problem of Mars and the much bigger problem of the structure of the universe. This situation was to end tragically in 1601 AD when Tycho died, in a manner that he would have appreciated. Kepler was appointed to succeed Tycho at the Imperial Court and he finally had access to Tycho's treasured data, data that would change the course of man's quest to understand the nature of the physical world.

In October 1604 AD while Kepler was observing a rare conjunction of Mars, Jupiter and Saturn a galactic supernova appeared in the sky and remained visible for 17 months. Within 32 years two new stars had appeared in the heavens, heavens which had been considered, since ancient times, to be pure and changeless. Kepler published his observations in 1606 AD as *De Stella Nova in Pede Serpentarii* (The New Star in the Foot of the Serpent Bearer). This brought him fame and recognition as an astronomer.

Kepler, with access to Tycho's incomparable collection of astronomical observations, resumed his study of the motion of the planet Mars. He started by assuming a circular orbit and thought that the problem would be simple as he was repeating the work of Copernicus but with more accurate data. He was wrong: it took him eight years. After extensive and laborious calculations he realised that the motion of Mars could not be explained with the Earth at the centre. The motion had to be referred to the position of the Sun. He concluded that there was 'one moving intelligence in the Sun that forces all round' with the planets close to the Sun moving faster than those further out. We see here that Kepler's thinking was still firmly Aristotelian with a belief in a 'prime mover', a concept familiar to Christian and Islamic medieval thinkers. Kepler grappled with these concepts for eight years but he could not reconcile them with the high precision of Tycho's data. In a bold step and after 900 pages of calculations, he decided to abandon the circular orbits.

Kepler's rejection of the circular orbits was the crucial departure from the dogma of the past, the step that Copernicus had been unable to take. The rejection of the circular orbits was forced on Kepler by the disagreement with Tycho's data. The disagreement was small; the path of Mars was just eight minutes of arc longer on one side of the Sun than on the other. But Kepler's faith in Tycho's data was absolute, and he was prepared to abandon his belief in the Platonic ideal of circular motion rather than question the data. Kepler realised that the planetary orbits are ellipses with the Sun at one focus. This observation swept the epicycles, deferents and equants out of astronomy. The pre-Keplerian dogma that permitted only circular orbits required that planetary motion be uniform, that is, a planet must traverse equal arcs in equal intervals of time. This was incompatible with elliptical orbits. But Kepler found an alternative form of uniformity. He showed that as a planet revolves in its orbit the line joining the planet and the Sun sweeps out equal areas of the ellipse in equal intervals of time. Hidden under this observation was a fundamental law of physics – the conservation of angular momentum, which was not discovered for another 75 years. Kepler published his work in 1609 AD in *New Astronomy with*

Commentaries on the Motion of Mars, a monumental work of 400 folio-sized pages in six parts. Kepler's style of presentation is very different from that of Copernicus or Tycho Brahe, both of whom presented the results of their work in a formal manner without any digressions. Kepler wrote informally, describing his motives, his hopes, his false starts, his frustrations and – finally – his triumph. In the *New Astronomy* Kepler sets forth his first two laws of planetary motion:

- Planets move round the Sun not in circles but in ellipses, the Sun being at one focus.
- A planet does not move with a uniform speed but moves such that a line drawn from the Sun to the planet sweeps out equal areas of the ellipse in equal time.

Kepler's mathematical skills were outstanding but he must have also been highly intuitive, he could not have discovered his second law just by analysis of the data. The *New Astronomy* gives arguments and proof of these two laws and it is full of other important suggestions, notably that the Earth attracts a stone just as a stone attracts the Earth, and that two bodies will attract each other if they are beyond the influence of a third body. We see here a glimmer of the theory of universal gravitation but this was not developed further. Kepler did, however, develop a theory of tides based on the attraction of the Earth's oceans by the Moon. Kepler also noted that a planet speeds up close to the Sun and slows down at a regular rate as it moves further from the Sun. This led him to suspect that the planets were kept in their orbits by a force which decreased with distance from the Sun. He actually considered a force that varies as the inverse square of distance but he did not consider gravitation, opting instead for a magnetic Sun, which to him was the obvious answer. Kepler was a generous man: in the *New Astronomy* he acknowledges his debt to Tycho and in particular to his data.

Kepler was not entirely satisfied with the regularities in the motion of the planets, which he had discovered; in particular he was interested in the possible relation between the period and the distance of a planet. This was the problem he had first considered in 1595 AD, in Graz. He realised that the problem was with the known distances of the planets. These were so inaccurate that they were practically useless for determining the relation he sought. He therefore decided to tackle a problem central to astronomy – measurement of the distances of celestial bodies. Aristarchus had first considered this problem in the third century BC and practically no progress had been made since. Kepler once again demonstrated his outstanding analytical skills by developing geometric methods to measure the distances of planets from the Sun. It took him another ten years to determine the distance to the five known planets, measured relative to the distance from the Earth to the Sun. The Earth–Sun distance was not measured accurately until the second half of the seventeenth century. But Tycho's data were of sufficient accuracy for Kepler to determine the relative distances to planets with enough precision to establish his third law of planetary motion. This he published in 1618 in his book *Harmonice Mundi* (*The Harmony of the World*). The Third Law states:

- The square of the period of revolution of a planet round the Sun is proportional to the cube of the average distance of the planet from the Sun.

Kepler called this Third Law the harmonic law because it represented to him the 'divine harmony of the world'. This law convinced him that the 'ultimate secret of the universe' would be found in an 'all embracing synthesis of geometry, music, astrology, astronomy and epistemology'. Kepler did not realise that his Third Law unambiguously suggested an inverse square law of attraction between the Sun and a planet. Newton made this discovery almost 75 years later. Nonetheless, Kepler's three laws of planetary motion revealed a solar system in which various parts were mathematically related.

Kepler was a complex and deeply mystical man; he had deep and profound faith in the harmony of the universe including a belief in the harmony between the universe and the individual. His skill in astrological predictions was much in demand. His first publication from Prague was *De Fundamentis Astrologiae Certioribus* (The More Reliable Base of Astrology) published in 1601 AD. In this he rejected the superstitious view that the stars guide the lives of human beings, but he continued to seek, in the events of his own life, a verification of the influence of heavenly bodies on human fate and behaviour. It is remarkable that his three laws of planetary motion, discoveries of the highest importance in arriving at an understanding of the physical nature of the universe, lie buried in deep mystical speculations. But his world-view was essentially Platonic, he was convinced that the arrangement of the universe must correspond to some abstract concept of beauty and harmony, and further, that this order must be mathematically expressible. To preserve the underlying simplicity of the complex nature of the observed universe he abandoned the Aristotelian geocentric model and the geometric world-view favoured by the Greeks. The natural order he discovered was not based on regular solids but on conic sections. The theory of conic sections was developed in late third century BC by Apollonios of Perga in Asia Minor (modern Turkey). A conic section, as the name implies, is a shape formed when a plane intersects a cone – a circle when the plane is perpendicular to the axis of the cone and an ellipse when the angle between the plane and the axis of the cone is greater than the semi-angle of the cone. A hyperbola and a parabola are also conic sections. To the Greeks the conic sections were abstract concepts to be studied as intellectual exercises. Kepler showed that these abstract concepts corresponded to reality in nature. Kepler was also the first natural philosopher to express laws of nature as mathematical equations. This expression was so successful that in the following centuries this became the only acceptable mode of expressing scientific ideas. Because of the universality of mathematics and mathematical symbols, scientific developments in different parts of the world were communicated and exchanged easily and more importantly, unambiguously.

In 1610 AD, a year before the publication of the *New Astronomy*, a close friend and admirer of Kepler brought him news of an Italian scientist who had built an instrument, based on a Dutch invention, with which he had observed valleys and mountains on the Moon, four moons revolving round the planet Jupiter and countless new stars in the sky. The instrument was the telescope and the Italian scientist was Galileo Galilei. In 1604 AD Kepler had applied the principles of optics to explain the process of vision, and he now used these principles to explain how a telescope works. Most intellectuals of the time viewed Galileo's findings with scepticism and ridicule, but Kepler acknowledged his accomplishments. Galileo did not return the compliment; although the two scientists had 'been in touch' Galileo completely ignored the epoch-making discoveries of Kepler!

Kepler died on 15 November 1630 in Regensburg, Germany. Sadly his grave was obliterated during the Thirty Years' War which tore apart Christianity into the Protestant and Roman Catholic churches.

3 Galileo

Gunpowder, an invention imported from China, proved immensely popular with the warring princes of fifteenth-century Europe. These princes were using gunpowder in their frequent wars, to hurl large projectiles at or over the walls of towns and cities they were attacking. By the middle of the sixteenth century the casting and boring of cannons had progressed to a stage where serious consideration had to be given to aiming and firing of guns. All over Europe gunners began to look at ways of increasing the range and aim of their artillery. But the path of the cannon ball made no sense within the context of Aristotelian doctrine. The Aristotelian laws of motion stated that the natural state of all 'earthly' objects was to be at rest. Motion away from the centre of the Earth was only possible with a 'mover' which had to be in contact with the object being moved. When the mover was removed the object should fall straight down to Earth. But cannon balls (or projectiles generally) did not fall straight down to Earth after they left the muzzle of the gun – they followed a curved path. Even the most ardent supporter of Aristotle could see that there was a flaw in the Aristotelian laws of motion. An alternative to the Aristotelian attempts to explain the motion of projectiles was the concept of the impressed force. According to this view, there is an incorporeal motive force that is imparted to the projectile, causing it to continue moving. This view was proposed in the seventh century by John Philoponus, by the eleventh-century Persian philosopher Avicenna and the twelfth-century Arab philosopher Abu al Barakat al-Baghdadi. The French philosopher and scientific theorist

Jean Buridan (1300–1358 AD) developed a new version of the impressed-force theory. He called the quality impressed on the projectile 'impetus'. The impetus is impressed by an external force but also accompanies any speed acquired naturally in free fall. A body commencing to fall gains certain speed in its 'first' motion when only heaviness and not impetus is acting. In the 'next' phase of motion the acquired speed adds impetus to the body's natural motion, and this continues indefinitely. The impetus was proportional to the speed of the projectile and matter contained in it. In addition he proposed (correctly) that the resistance of air progressively reduced the impetus. The moved object was still assumed to move in a straight line, which was the only path allowed on Earth. The theory of impetus was compatible with Aristotelian doctrine as the impetus could be identified with the Aristotelian 'qualities' of a body (heat, cold, dryness and moisture) – impetus is the quality that provides motion. According to this theory motion causes impetus to appear in a body; the faster a body moves the more impetus it possesses. This is why a dropped object accelerates as it falls. The impetus theory was widely accepted for almost 200 years after it was published.

In the last decades of the fourteenth century the study of Euclid's *Elements* particularly his ratio theory, provided the means of expressing mathematically various relations of the Aristotelian qualities associated with moving bodies. This quantification of qualities, the so-called latitude of forms, was a topic of intense study and discussion at the University of Paris and at Merton College, Oxford. The Oxford scholars pondered the philosophical problem of how to describe the changes that occur in a body when the Aristotelian qualities of a body increase or decrease. The Aristotelian qualities were assigned an intensity and extension, which were represented by the height and base, respectively, of a geometric figure. The area of the figure was then considered to represent the quantity of the quality. In the important case of motion (a quality) of a body, the intensity is its speed or velocity, and extension the time of travel. The area of the figure was taken to represent the distance covered by the body. Uniformly accelerated motion starting with zero velocity was defined by a triangle. The Merton school showed that the total quantity of accelerated motion is equal to the quantity of uniform motion achieved at the speed attained halfway through the accelerated motion. The distance travelled was equal to the area of the triangle and therefore proportional to the square of the travel time ($s \propto t^2$). This was known as the Merton Law. Over 200 years later Galileo was to derive this relation geometrically and prove it experimentally. It is possible that this late fourteenth-century attempt to quantify Aristotle's qualities influenced Galileo's foundation of the science of mechanics and the foundation of coordinate geometry, in the seventeenth century.

Towards the end of the fifteenth century the great Italian artist, scientist and engineer Leonardo da Vinci (1452–1519 AD) showed that a cannon ball travels along a curved path. With or without impetus this was a violation of Aristotelian doctrine. The doctrine stated that there were two types of movements – natural and violent or forced. The straight-line path of the cannon ball as it left the muzzle of the cannon was allowed on Earth as it was earthy and degenerate. But the curve that followed was 'natural' or celestial and had no place on Earth! Towards the middle of the sixteenth century Giovanni Benedetti took up the problem of cannon balls and falling objects. Benedetti was a student of Niccolo Fontana Tartaglia, a professor of mathematics at the University of Venice, who had suffered a sabre blow during the French sack of Brescia (1512 AD) which cleaved his jaw and palate. The resulting speech impediment earned him the nickname Tartaglia (stammerer), which he adopted. Tartaglia was mainly interested in gunnery and had published a book, which showed that the entire path of a cannon ball was curved and the maximum range was obtained if the cannon was fired at an elevation of 45 degrees. Benedetti decided to test the Aristotelian statement that the speed of fall of a body was related to its weight. He dropped two bodies of equal weight tied together with a thin thread. He reasoned that the joined bodies should fall faster than the individual bodies. This he did not find. He also showed that regardless of their weight, all bodies of same material fall with the same

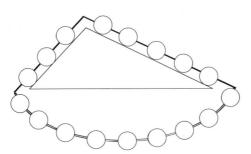

FIGURE 3.1 Stevin's 'chain of spheres' on a triangle. If the spheres hanging below the triangle are removed the chain will stay on the triangle because the downward force on the two sides is equal. This force is proportional to the total weight of the spheres and the angle (from the vertical) of the two sides.

speed. This was not universally accepted at the time and Benedetti had to reply to vigorous attacks against this proposition. Benedetti also decided to take a critical look at the impetus theory and the concept of circular motion. He rejected the Aristotelian idea that bodies went to some 'preferred place'. He noticed that a stone swung round in a circle at the end of a rope flew off in a straight line (along the tangent to the circle) when released. He reasoned that this showed that impetus could cause bodies to move along both straight and curved paths. Benedetti was one of the principal thinkers in the philosophical revolution that was sweeping through Europe in the sixteenth century. Unfortunately he failed to receive the recognition he deserved; his work was plagiarised and popularised by others.

By the late sixteenth century the qualitative world of Aristotle based on abstract concepts of 'elements and qualities' was under sustained attack. In 1586 AD the Dutch accountant-turned-scientist Simon Stevin (1548–1620 AD) published his *De Beghinselen der Weeghconst* (Statics and Hydrostatics) in which he gave the first demonstration of the resolution of forces, now known as the parallelogram of forces. He showed that a necklace of metal spheres laid over a triangle, apex up, would hang on the triangle (Figure 3.1). If all the spheres hanging below the triangle were taken away the remaining spheres would stay at rest even if one side had more spheres than the other. This is because the downward force on the two sides is equal and is proportional to the angle of the two sides of the triangle (from the vertical) and the total weight on each side.

In 1586 AD Stevin also published a short report in which he showed that two lead spheres, one ten times as heavy as the other, took same time to fall a distance of about 9 metres. This experiment was similar to the experiments on falling bodies done by Benedetti and preceded by 18 years Galileo's work on falling bodies, but it received little attention at the time. But this phenomenon – that is, in the absence of resistance, all bodies irrespective of their weight or composition fall at the same rate – has puzzled physicists ever since. This was demonstrated graphically by the first American astronaut to land on Moon, by dropping a golf ball and a feather. In the absence of air resistance the two objects took same time to fall down to Moon's surface. In the twentieth century Einstein was to use the 'free fall' of bodies to formulate his theory of relativity (see Chapter 5, Einstein). A space satellite experiment has been planned for early in the twenty-first century to test to very high precision this 'principle of free fall' (see Chapter 6, Dicke).

MECHANICAL PHILOSOPHY

In 1637 AD the French philosopher scientist Descartes published *Discours de la méthode* (Discourse on Method), one of the first important philosophical works not published in Latin. Descartes said that he wrote in French so that all men and *women* who had good sense could read his works and learn to use their reason and think for themselves. Descartes' *Method* revolutionized European thought just as the Aristotelian view had done before him. René Descartes (1596–1650 AD) was born in La Haye in the south of France. His mother died when he was one year old and his maternal grandmother raised him. In 1606 AD he was sent to the Jesuit college at La Flèche where young gentlemen were trained for careers in military engineering, the judiciary and government administration. Here he was instructed in classics, sciences, mathematics and metaphysics and also acting, music, poetry, dancing, riding and fencing. Young Descartes, instead of following a legal and political career as he was probably expected to do, decided to go to Breda in the Netherlands in 1618 to study military architecture. At Breda, he was encouraged to study science and mathematics. Between

1619 and 1628 he travelled extensively in northern and southern Europe studying, as he said, 'the book of the world'. By 1620 he had conceived a universal method of deductive logic, applicable to all sciences. In 1624 the Parliament of Paris passed a decree forbidding attacks on Aristotle on pain of death. Descartes decided to leave France for the Netherlands to be free from interference by the ecclesiastical authority. This he did in 1628 and did not return to France for next 16 years.

Descartes wrote perhaps the most famous sentence in the history of philosophy – *cogito, ergo sum* (I think, therefore I am). Starting with this startling sentence, Descartes proceeded to build a theory of the nature of man and the universe. Descartes rejected the evidence of the senses. He believed that the commonly accepted knowledge was doubtful because of the subjective nature of the senses. In three essays forming part of the *Method* he illustrates ways of utilising reason in search of truth in the sciences. He gives four rules for reasoning: (1) Accept nothing that is not self-evident. (2) Divide problems into their simplest forms. (3) Solve problems by proceeding from simple to complex. (4) Recheck the reasoning. These rules are an application of mathematical procedure to science. Descartes also believed that the entire material universe could be explained in terms of mathematical physics, and in *Geometry* (one of the three parts of the *Method*) he gave an exposition of analytical geometry. This is a method of representing geometric figures with algebraic equations, and it made many previously unsolvable problems solvable. This application of algebraic methods to geometry was perhaps the greatest single step in the seventeenth century in the progress of the exact sciences. Descartes saw a curve as a path of a moving point, the point being the intersection of two lines which are always parallel to two fixed lines which are at right angles to each other. The motion of a body can be described by list of positions and corresponding times. A continuous description would require a mathematical equation expressing the position from the two fixed lines in terms of time. This has become known as the 'Cartesian coordinate' system. An example of a simple two-dimensional Cartesian coordinate system is shown in Figure 3.2. A two-dimensional coordinate system is all that is required to describe the

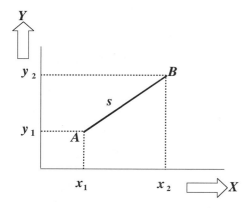

FIGURE 3.2 A two-dimensional coordinate system. The length of the line AB is given by Pythagoras's theorem. This coordinate system is used to define the position of a point on a surface like a plane paper.

position or motion of bodies on a flat surface. For example, the position of point A can be represented by two coordinates (x_1, y_1), and similarly B by (x_2, y_2). The length of the line AB is given by

$$s^2 = x^2 + y^2$$

where $x = x_2 - x_1$ and $y = y_2 - y_1$. The length of the line is obtained by Pythagoras's theorem. To follow points or bodies in the real world, for example a falling body, it is necessary to have a three-dimensional coordinate system, as the real world is three-dimensional. A three-dimensional coordinate system is obtained by introducing a third axis (line) perpendicular to the surface defined by the two-dimensional coordinate system; this is shown in Figure 3.3. The length of the line AB is now given by

$$s^2 = x^2 + y^2 + z^2$$

which is a three-dimensional generalization of Pythagoras's theorem.

Descartes had thus introduced geometry into the description of motion. This concept has seen innumerable developments and is now adopted in the sciences in general. Its most familiar development is the 'graph'. Descartes also introduced a number of mathematical conventions, which made algebraic notation much clearer than it had been before. The new analytical geometry did away with the cumbersome

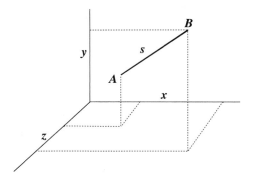

FIGURE 3.3 A three-dimensional coordinate system (this is similar to the corner of a room). This coordinate system is required to define the position of a body in the 'real' world. The length of the line AB in this case is given by a 'generalised' application of Pythagoras's theorem.

geometric drawings used by Kepler to derive his laws of planetary motion and permitted all forms of motion to be analysed algebraically. The equations for the trajectory of any projectile could be written and then manipulated mathematically to show what would happen to the projectile under altered conditions such as increase in propulsion or weight. The method was particularly useful for applying laws of motion investigated on the Earth to the motion of planets, which are, of course, out of reach.

Descartes also applied the 'Cartesian logic' to the universe. In 1640 he wrote *The Principles of Philosophy* in which he described a universe which is an indefinitely large continuum (*plenum*) filled with infinitely divisible matter. Movement in any part of this continuum was assumed to cause the rest of the matter to respond. The continuum is separated into the subtle matter of space and the denser matter of 'material' bodies. The subtle matter is set in motion by God and this matter imparts motion to material bodies. The rotation of the Sun was assumed to be caused by the spinning subtle matter creating a kind of whirlpool. Planets were assumed to be caught in this vortex and carried round the Sun. Minor vortices were assumed to cause terrestrial motions. The action of gravity was explained as the effect of the particles of the continuum, set in motion by the Sun, exerting a force on terrestrial objects and make them fall to Earth. Descartes followed Benedetti in thinking that without vortices planets would be flung out in straight lines away from their orbits. His was a mechanical universe, in which nothing happened except as a result of impact between particles. Matter had a tendency to do nothing until it suffered an impact; a object in motion moved as it did because its inert state had been altered by impact. In his assumption of inertness of matter Descartes had made the greatest advance; here was the germ of the concept of *inertia* that was finally teased out by Newton. Descartes was puzzled by the question that had puzzled both Copernicus and Kepler. If a body moves, relative to what does it move? Copernicus and Kepler had concluded that a huge, rigid frame on which the stars were fixed bound the universe, with the Sun at its centre. All motion was relative to this frame. Descartes seems to have concluded that there was no such frame and that all motion was relative. But he did not publish this work, fearing the wrath of the Inquisition, which had condemned Galileo (for teaching that the Earth moves). He did publish his ideas in 1644 in *Principia Philosophicae,* but these are logically inconsistent. Descartes argued that a moving body moves only relative to another body, chosen as a reference. Since the choice of the reference body is not restricted, a moving body can have a number of different references and therefore a number of different motions, which are all relative. However, he did allow a moving body 'one absolute motion', that relative to the matter immediately adjacent to the body itself. Thus, although the Earth is carried around the Sun in a huge vortex, it does *not* move relative to the matter in its immediate vicinity! This he thought would let him off the Inquisitorial hook.

Based on the notion of impact between particles, Descartes proposed two laws of motion[1]:

- If two bodies have equal mass and velocity before collision then both will be reflected by collision and will retain the speed they had before collision.
- If two bodies have unequal masses, then upon collision the lighter body will be reflected and its new velocity will be equal to that of the heavier body. The velocity of the heavier body will remain unchanged.

These Cartesian 'laws' were not derived (or verified) from observations of colliding bodies. An experiment would have revealed the 'defect' in these laws, namely the problem of discontinuity. Imagine two masses M and m with velocities V and v respectively. If $M = m$ and $V = v$ then after collision of the two masses $V = -V$ and $v = -v$, where the negative sign indicates reflection. This is the first Cartesian law of motion. If M is greater than m then after collision $v = -V$ and $V = V$, according to the second Cartesian law of motion. But if the mass M is now gradually reduced to approach the value of m (that is a situation where the first law would apply) then at some point $V = -V$ and this is only possible if $V = 0$. It was more than 50 years before this contradiction was recognised by Leibniz.

The Cartesian material world is a deterministic world, because the amount of motion is conserved according to the laws of nature. If the speed, the amount of motion and position of all the whirling portions of the matter in the universe are known at any one time then their descriptions at any subsequent time could be deduced from the laws of motion. But the theory of vortices failed to explain a whole host of known phenomena, including Kepler's laws of planetary motion. Unfortunately it also became very popular and an elaborate system of physics and cosmology was built on it. In France it survived until the middle of the eighteenth century. Gradually the numerous errors that it contained were exposed and the theory was eventually made untenable by the work of Newton. This is one of those sad examples of the many blind alleys of science.

Descartes wanted to explain all phenomena of nature in terms of matter and motion. This view, known as mechanical philosophy, came to dominate seventeenth-century science. Descartes was concerned with the problems of the foundation of science and not with the solutions of specific problems, and he belittled the work of Galileo because it did not include an exposition of the fundamental concepts with which Galileo was dealing, namely force, matter, space and time. The Cartesian philosophers considered the neglect of causes and universality to be irremediable and fatal flaws in Galileo's work. These philosophers demanded an impossible certainty and applicability of science and this claim was to become the keynote of mechanical philosophy. Fortunately for science, Galileo steered away from mechanical philosophy and went on to lay the foundation of experimental physics and to become an acute exponent of natural laws. Thanks to him, by the end of the seventeenth century the Cartesian rationalistic physics had been abandoned for the empirical results of observation and experience.

BIRTH OF EXPERIMENTAL PHYSICS

Galileo Galilei (1564–1642 AD) created the modern idea of experiment. He was also the first man to realise that mathematics and physics, previously kept in separate compartments, should join forces. He, more than any other natural philosopher of the seventeenth century, was responsible for replacing Aristotle's logico-speculative approach with mathematical rationalism; he emphasised this in his insistence that the "Book of Nature . . . is written in mathematical characters". His long life was blessed by unparalleled intellectual activity and blighted by ignorance and intolerance. Galileo was born in Pisa on 15 February 1564. His early education was at a monastery near Florence, where his family had moved. In 1581 AD, at the age of 17, young Galileo went to the University of Pisa to study medicine, but he soon tired of the doctrinal texts of Aristotle and other Greek works. During his stay at Pisa he overheard a lecture on geometry being given to the pages of the Tuscan court, which so fired his interest in mathematics and physics that he started taking lessons in these subjects. While praying at the cathedral in Pisa he is supposed to have noticed the swinging chandelier and noted that the lamp always required the same amount of time to complete an oscillation, no matter how large the range of the swing[2]. Later in life he was to conduct experiments which established that the period of oscillation of a pendulum of a fixed length is independent of the amplitude, the weight and the nature of the suspended body. This suggested to him that the pendulum could be applied to regulate clocks.

In 1585 Galileo left the University of Pisa without taking a degree because his family could no longer afford the university fee. He left with a reputation for extraordinary talent in mathematics and also a reputation for being both an independent spirit and an iconoclast. Galileo returned to Florence to teach at the Florentine Academy. There his thesis on the hydrostatic balance, published a year later, made his name known throughout Italy. Around this time he also began to reflect on the nature of measurement and its vital role in science. Galileo regarded demonstration unsupported by experience to be 'the world on paper' while actual measurements were 'the real world'. An essay on the centre of gravity of solids, published in 1589, won for him the honourable but not so remunerative post of lecturer in mathematics at the University of Pisa. Here he started investigation of accelerated motion. He was to show through experiments that a number of Aristotle's conclusions, which for 2000 years had been held to be authoritative, were actually false. He conducted experiments with balls rolled down an inclined plane, effectively 'slowing down the free fall of bodies' so that the fall could be timed. Through his experiments he arrived at a clear understanding of acceleration as well as of the concept of inertia. At Pisa, Galileo laid the foundation of the science of kinematics (a branch of mechanics that deals with the description of motion of bodies). He (and others) attempted to formulate an explanation of the cause of motion (the science of dynamics) but this was not successful; the task was to be left to Newton. Tradition has it that he disproved the Aristotelian doctrine of motion, that the speed of falling bodies is proportional to their weight, by dropping two different objects simultaneously from the Leaning Tower of Pisa. Sadly, there is no evidence that this experiment was actually performed. Galileo has not mentioned it in his well-kept notes and there is no record of this experiment in the archives of the University. This 'Tower of Pisa experiment' was, however, performed in 1993 by members of a European collaboration considering a satellite experiment to test the theory of free fall (see Chapter 6, Dicke). At Pisa Galileo wrote a thesis, *De motu*, which was an improvement on the contemporary discussion of motion. Using more mathematics than was customary in dealing with motion, he refuted many 'received' opinions. He refuted the Aristotelian notion that the medium collaborated in the motion of a body moving through it. He also asserted that the vertical motion of a body, up or down, could be described by weight alone. However, he retained the concepts of 'lightness', 'heaviness' and the proximate cause. Because the conclusions reached in *De motu* had not been established by experiments, Galileo felt that the arguments advanced were not sufficiently convincing to be published. But by contradicting Aristotle he had managed to offend professors of philosophy and it was unlikely that his contract at the University would have been extended.

In 1592 Galileo was appointed Professor of Mathematics at the prestigious University of Padua. Galileo was 28 and he was to spend the next 18 years at Padua completing the bulk of his work on mechanics. Here he proved theoretically (around 1604) that falling bodies obey the law of uniformly accelerated motion (that is, motion in which the speed of a body increases or decreases uniformly in time). He also formulated the law of parabolic fall. He was to write later, in his *Dialogues Concerning Two New Sciences* (1638):

> It has been observed that missiles and projectiles describe a curved path of some sort; however, no one has pointed out the fact that this path is a parabola.

He had deduced that the motion of a projectile was the consequence of simultaneous and independent inertial motion in the horizontal direction and falling motion in the vertical direction.

Galileo gives a full exposition of his experimental and theoretical work on motion in the *Two New Sciences*, completed in 1634. He wrote that 'there is, in nature, perhaps nothing older than motion' about which 'books written by philosophers are neither few nor small'. But most of these writings were qualitative and even speculative. In the *Two New Sciences* Galileo proposes that any science of motion should be mathematical and experimental. Galileo was to write that he had 'discovered by experiment some properties...'. The book is divided into four parts or *days,* the first two days dealing with the strength of materials and nature and properties of fluids, the third dealing with motion and the fourth with projectiles. The third day is divided into two parts; the first part deals with uniform motion and the second with naturally accelerated motion. Galileo retained the rather handy distinction, made by Aristotle, between natural and forced motion. Natural motion was motion in a vertical line near the surface of the Earth, that is free fall. This distinction has now been eliminated by reducing natural motion to motion under 'the force of gravity'. In the *Two New Sciences* Galileo defines uniform motion as follows[3]:

> By steady or uniform motion, I mean one in which the distances traversed by a moving body during *any* equal interval of time, are themselves equal.

Galileo is being very precise by inserting the word 'any' – the distances must be equal for all equal intervals of time and not equal in an average sense over the total distance. From this definition Galileo derives a number of axioms before defining velocity as $v = s/t$ (in modern notation), where s is the distance traversed in time t. Galileo then proceeds to investigate motion of accelerated bodies. He defines uniformly accelerated motion as when[3]

> A body acquires equal increments of speed during any equal intervals of time

This is equivalent to $v \propto t$, a formula taught in the introductory course in physics. It is worth noting that this was an intuitive leap by Galileo: he had defined uniformly accelerated motion without any experimental evidence. He then proceeded to 'prove' this definition deducing mathematically the consequences of such motion. He showed that for uniformly accelerated motion the distance traversed increases as the square of the elapsed time ($s \propto t^2$) and the velocity acquired in free fall from different heights is proportional to the square root of the height ($v \propto \sqrt{h}$). He goes on to give logical arguments to prove continuity of motion, which was not at all obvious at the time.

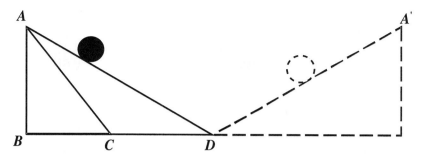

FIGURE 3.4 The inclined planes experiment of Galileo. A heavy body (to reduce the effect of air resistance) released from A can either fall vertically down to B or can be made to roll down an inclined plane AC or AD. Whichever path the body takes it will reach points B, C and D with equal speed. Also, if there is no 'discontinuity' at D, the body will move along the plane DA' to a height equal to the height from which it is released.

He applied his formulae to motion down an inclined plane and showed that in the absence of all resistance or opposition the speed acquired by any body moving down planes of different inclination is equal when the height of these planes is equal. That is, a heavy (to reduce the resistance of air) and perfectly round ball descending along the lines AB, AC, and AD (Figure 3.4) would reach the terminal points B, C and D with equal speed. He also asserts that a body descending a plane will acquire 'momentum' (this should really be energy) sufficient to carry the body back to the same height. That is, a ball descending from point A along the line AD will be carried to the point A' that is at the same height as A. Galileo is careful to point out that there is a discontinuity at the point where the planes meet (i.e. D) and that this will present an obstacle to the descending ball. But he proves his assertion by a pendulum experiment, as shown in Figure 3.5. If the bob is pulled to position C and released it will descend along the arc CB to reach the lowest point B and then continue along the arc BA to reach point A where the vertical distance CB will equal AB. Galileo was careful to stress the difference between motion along a curved path and motion along a straight path. The principle of conservation of energy (kinetic plus potential) is implicit in these

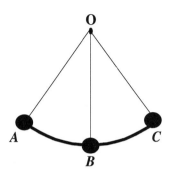

FIGURE 3.5 Pendulum experiment to prove the inclined planes experiment. A bob released at C will proceed along the arc CB and continue along the arc BA. The height CB is equal to the height AB. In moving from C to B the bob loses potential energy but gains kinetic energy which is equal to the potential energy it has lost. This energy is sufficient to take the bob to its original height, thus converting the kinetic energy to potential energy it had at C. Kinetic and potential energy are described more fully in Newton – Chapter 4.

experiments, but it was some time before this was realised (see Chapter 4, Newton).

Starting with this simple experiment Galileo made the great intellectual leap and proposed the concept of inertia. He argued that if the height of the point A' (Figure 3.4) was reduced then a ball rolling down the inclined plane AD would travel further and further as the height of the second plane is reduced. When the plane was horizontal the ball would continue to roll indefinitely unless stopped by any other means like friction or an opposing force. (But Galileo got it slightly wrong: he thought that the indefinite motion would be circular, round the Earth. This was because a circle is a perfect Platonic figure. Galileo's inertia was circular. Fortunately the difference between a straight line and a curve was too small in Galileo's experiments to affect his analysis. Galileo's classical education clearly had influenced his thinking.) The principle of inertia is encapsulated here – a body will continue in its state of rest or motion in the absence of application of an altering force.

Galileo thus showed that force was not necessary for motion, but only necessary to bring about a change in motion. Galileo had also demonstrated that no proximate cause was necessary for motion. At a stroke he had demolished the Aristotelian notions of motion and causes of motion. Aristotelian doctrine maintained that a body would move as long as an agent moved it. Galileo insisted that motion is preserved and requires no agent. A body set in motion on a perfectly smooth horizontal surface will continue to move forever if the surface has no limit. In the *Two New Sciences* Galileo created the new science of mechanics, in which the concept of motion is radically different from the concept that had prevailed for the previous 20 centuries. The new concept of motion that he proposed is substantially our current conception.

Around 1609 Galileo received details of the basic principle of the telescope, invented by a Flemish inventor[4]. He quickly calculated the ideal shape and placement of lenses, ground and polished the necessary lenses himself and constructed a telescope that had a magnification of about 10. The power of this instrument for military and commercial purposes was immediately recognised by the Venetian Senate who provided funding for a larger telescope and also more than doubled Galileo's salary. Galileo had devised a method for checking the curvature of the lenses and the quality of image of his telescopes was considerably superior to that achieved elsewhere. The new telescope Galileo built had an aperture of about 5 centimetres and a magnification of about 20, but instead of using it to observe ships at sea, Galileo turned it towards the sky and the world of observational astronomy was altered forever. Galileo's interest in astronomy had been kindled in 1604 when, on 8 October, astronomers on the lookout for the predicted conjunction of Jupiter and Mars (of considerable astrological significance) in the constellation of Sagittarius noticed a new star instead. The new star was very bright, approaching the brightness of Venus. This was the same 'new star' which Kepler had also observed. Galileo was informed of the new star around 15 October and he immediately started careful observations of its brightness and position. He also wrote to other cities, especially Verona, to obtain similar data. The brightness of the star excited much public curiosity and Galileo delivered three public lectures on it to very large audiences. In these lectures he explained the nature and application of parallactic reasoning to measure distances and refuted the Aristotelian theory that the new stars and comets were sub-lunar phenomena in the supposed region of fire above the air and below the Moon. Lack of parallactic displacement suggested to Galileo and other astronomers at Padua and Verona that the new star was at least at the distance of the outer planets and possibly as far out as the fixed stars. Parallactic displacements were familiar to land surveyors, in judging the distance and height of remote objects. But the application of these terrestrial procedures to celestial phenomena and to reach conclusions from these measurements was a direct challenge to Aristotelian and Christian doctrines. Galileo's lectures were immediately challenged in debates in the university and Galileo was identified as the leader in this battle against the Aristotelian philosophy.

But the 'Aristotelian controversy' generated by the 1604 nova paled into insignificance when Galileo revealed the discoveries he had made with his telescope. In late 1609 and early 1610, Galileo announced a series of astronomical discoveries which stunned the scientific world. He had observed mountains and craters on the Moon and also sea-like (maria) regions, planets appeared disk-shaped but stars were point-like, Venus showed phases like the Moon, and Jupiter had four moons which were in the same plane and revolved around the planet – a miniature Copernican system. Galileo named these moons *Sidera Medicea* (Medicean Stars, now known as the Galilean satellites of Jupiter) in honour of his future employer (and former pupil) Cosimo II, Grand Duke of Tuscany. Within two years of this discovery, he had created accurate tables of the period of revolution of Jupiter's satellites and proposed their frequent eclipses as a means of determining longitudes on land and sea. The idea, though ingenious, proved of little use at sea because of the difficulty of making observations of stellar or semi-stellar objects. Galileo also noticed extensions beyond the disk of Saturn but failed to interpret their significance. His observations of the star cluster known as the Pleiades (Figure 3.6) produced another surprise: apart from the six naked-eye stars he observed 36 other stars, 'none much more than half a degree away from the six'. Similarly, the

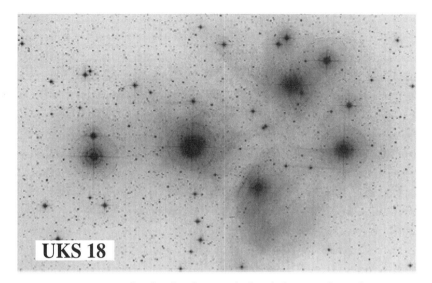

FIGURE 3.6 The Pleiades photographed with the United Kingdom Schmidt telescope of the Anglo-Australian Observatory. The Pleiades contain about 4000 stars, but with the unaided eye Galileo would have seen only the six brightest stars. The 'fuzz' in the picture is light emitted by gas which has been ionized by radiation from the bright stars. (Image by D. Malin, Anglo-Australian Observatory.)

Milky Way appeared to have far more stars than those visible to the naked eye. This was the first observational evidence against the 'thin-starry-sphere universe' of Aristotle; the faint stars were either intrinsically faint or they were more distant. The way had been opened for a 'thick-starry-sphere' universe. The observations of the phases of Venus, which he made in 1610, confirmed for Galileo the correctness of the Copernican astronomy. Neither Aristotelian nor Ptolemaic astronomy could account for the observed phases. The scheme proposed by Tycho could account for these observed phases but Galileo rejected this on the grounds of physics. To him a stationary Earth with an orbiting Sun carrying all the other planets with it was patently absurd. In 1610 Galileo published his astronomical discoveries in *Siderius Nuncias* (The Starry Messenger). These 24 pages were to lead to his downfall.

In recognition of his telescopic discoveries the Venetian Senate granted Galileo lifetime tenure as professor at the University of Padua. But in the summer of 1610 he left this position to return to Florence to become the 'first philosopher and mathematician' to the Grand Duke of Tuscany, and the chief mathematician of the University of Pisa. This appointment had no teaching duties and Galileo was able to devote more time to research. In 1611 he was elected (the sixth) member of *Accademia dei Lincei* in Rome. The *dei Lincei* was perhaps the first scientific society in the world (see Chapter 4, Newton). In Rome he also demonstrated his telescope at the Pontifical Court. Encouraged by the positive reception accorded to him he ventured, in three papers on sunspots, published in Rome in 1613, to take a firm stance on the Copernican theory. The movement of spots across the face of the Sun, Galileo maintained, proved the rotation of the Sun and the revolution of the Earth. Galileo asserted that this proved that Copernicus was right and Ptolemy wrong. Galileo wrote in Italian, in which he was an acknowledged master of style, and his publications were easily accessible and popular in Italy. His fame spread far beyond the mercantile world of Venice but it also sharpened the opposition to his teachings by the Aristotelian academics and the Dominican preachers. In particular his observations of the Copernican world of Jupiter brought into sharp focus the contemporary discussion of the 'plurality of worlds' started by Bruno. Deeper, but not often expressed, was the philosophic fear of the infinite universe, which Bruno had also suggested and which was implied by the large number of faint stars observed by Galileo. Galileo was secretly denounced to the Inquisition for blasphemy. Shocked, Galileo wrote to the Grand Duchess Cristina (mother of Cosimo II, Medici) and the Roman authorities reminding them of the long-standing practice of the Church to interpret the Scripture allegorically whenever it came into conflict with scientific truth. He even went to Rome to plead with the authorities to leave the way open for change and scientific progress. Remarkably, a number of ecclesiastical authorities were on his side. It is worth emphasising that Galileo was a devout Catholic and did not question the spiritual authority of the Catholic

Church, but he did rattle the Aristotelian skeleton in the Church's cupboard. Pope Paul V summoned Roberto Cardinal Bellarmino, the pre-eminent Jesuit theologian of the church (who had served as inquisitor in the trial of Giordano Bruno), to reconsider the Copernican model of the solar system and whether it should be condemned as heretical. Bellarmino, more than most, was able to appreciate the importance of the new theories and discoveries of Galileo, having studied astronomy himself at Florence. He also highly respected Galileo and his achievements, but he could not accept Galileo's insistence that the Copernican model was a reality and not a hypothesis. He chose the safety of the time-honoured belief that 'mathematical hypotheses have nothing to do with physical reality'. He saw dangers in the academic split that the discoveries of Galileo and his writings had caused. He felt that this would undermine the fight of the Catholic Church against Protestantism. At the request of Pope Paul V, a panel of 11 theologians, drawn from the cardinals of the Holy Office, voted on the Copernican doctrine in February 1616. The unanimous verdict was that the doctrine was 'false and erroneous'. A decree was issued on 5 March 1616, putting the works of Copernicus on the proscribed list – 'suspended till corrected'. Out of personal consideration, Cardinal Bellarmino granted Galileo an audience a few days before the decree. Galileo was informed of the decree and was warned that henceforth he should 'neither hold nor defend' the doctrine, although it could be discussed as a 'mathematical supposition'.

Disappointed, Galileo returned to his studies in Florence. In 1623 AD he wrote *Saggiatore* . . . (Assayer . . .) in reply to a pamphlet on the nature of comets. The 'Assayer' is a brilliant polemic on physical reality and the exposition of the new scientific method. In it he distinguishes between the primary properties (such as a measurement) of matter and the secondary characteristics. The book was dedicated to the new pope, Urban VIII, who had been a long-time friend and like Galileo an alumnus of the University of Pisa. Pope Urban received the dedication enthusiastically. Encouraged, Galileo travelled to Rome to urge the papal authorities to revoke the decree of 1616. This he did not get but he did obtain permission from the Pope to write about the 'systems of the world', both Ptolemaic and Copernican, provided they were discussed noncommittally. He was also told that the book should come to the conclusion that 'man cannot presume to know how the world is really made because God could have brought about the same effect in ways unimagined by him, and man must not restrict God's omnipotence'. The Pontiff dictated the conclusion to Galileo and the head censor confirmed these instructions in writing.

Galileo returned to Florence and spent the next seven years working on his great book *Dialogo sopra i due massimi sistemi del mondo, tolemaico e copernicano* (Dialogue Concerning Two Chief World Systems – Ptolemaic and Copernican). The main themes of the *Dialogue* are Galileo's conception of the physical relativity of motion and his arguments for the motion of the Earth based on the annual variation of paths of the sunspots. Galileo also argued that as the Earth turned everything on it turned with it, so that falling objects move east with Earth. He reasoned that the two components of motion, that is, the downward motion of the object and the rotation of Earth, resolved to cause the object to reach a spot vertically below the release point. He compared this with dropping an object from the top of a ship's mast. It would hit the deck because both the ship and the object were travelling together. This explained the problem Copernicus had not been able to resolve: why falling objects do not fall to the ground to the west of their starting point. Galileo's argument also destroyed the Aristotelian separation and independence of violent and natural motions; an object's movement is determined by the composition of both motions. Apart from the discussion of the relative position of the Earth and Sun and their motion, the *Dialogue* attempts to present the doctrine of uniformity in the working of the material world. The doctrine of uniformity – every cause has an effect – seems self-evident to us now but in the Aristotelian concept of the universe, this was not so. According to this view the events in the supra-lunary sphere – the celestial spheres – were very different from the earthly happenings. Galileo was in no strong position to discuss celestial physics but he throws out a strong hint that it could be discussed in terrestrial terms, thus anticipating the theory of universal gravitation.

The *Dialogue* came out in 1632 with the full and complete imprimatur of the censors and it was an instant success, being greeted as a literary and philosophical masterpiece. The *Dialogue* is presented as a debate between three persons: an advocate of Copernican doctrine called Salviati (a thinly disguised alter ego who spoke Galileo's own mind), a pompous follower of Aristotle and Ptolemy called Simplicio, and Sagredo, an intelligent and receptive man of means who typically agreed with Salviati. The pro-Church view is put into the mouth of Simplicio (a pun on the word 'simpleton'). The pro-Copernican argument was brilliant but somewhat cruel (Galileo does not miss an opportunity to run down those who espoused Aristotelian views). Also, the Copernican argument goes far beyond Copernicus and totally rejects the Aristotelian 'fixed crystal sphere of stars'. The strength of the argument made a mockery of the prescribed conclusion, which Galileo had dutifully inserted, at the end of the book. The Catholic Church was 'not amused' and Pope Urban VIII unleashed the full force of the Inquisition. Despite pleading ill health and old age, Galileo was summoned to Rome in February 1633 to stand trial. He was found guilty of having 'held and taught' Copernican doctrine. Galileo, aware of the fate of Bruno, recited a humiliating confession that he 'abjured, cursed and detested his past errors'. He was sentenced to life imprisonment. Significantly three cardinals, of the ten cardinals of the Inquisition, refused to sign the sentence. The sentence of life imprisonment was immediately commuted to house arrest and seclusion. The Pope also agreed to release Galileo into the custody of Archbishop Ascanio Piccolomini, of Siena, a long-standing friend and sympathiser of Galileo. The sentence of house arrest remained in effect throughout the last eight years of his life. The *Dialogue* was placed on the Index of prohibited books, where it remained together with the works of Copernicus, until 1822.

Galileo was badly shaken by the trial but in the company of the cultured and gracious Piccolomini he soon regained his composure. Piccolomini, son of a distinguished family of scholars and a man who had himself studied mathematics, encouraged him to put behind him his months of anxiety and suffering and turn his thoughts to scientific pursuits. It was at Siena that Galileo began work on the *Dialogues Concerning Two New Sciences.* The first part of the 'Third Day', the section on local motion and the most celebrated part, was written at Siena. Towards the end of 1633 he was allowed to return to his little estate at Arcetri near Florence, although he was still under house arrest and was not allowed to go even to nearby Florence for medical attention. Galileo's prodigious mental activity continued undiminished to the last. In 1634 he completed the *Two New Sciences*, his greatest scientific achievement, published in Leiden, Holland, in 1638. In this work he emphasises the role of experimentation although he was aware that experiments alone did not constitute science. While he did not discover the first principles of dynamics (these were left to Newton) his scientific method was entirely modern. He made his last astronomical discovery, that of the diurnal and monthly libration of the Moon (its wobble from side to side), only a few months before he went blind. This handicap did not diminish his intellectual vigour, and he continued his scientific correspondence. He also worked out the application of the pendulum to regulate clocks, which the Dutch scientist Christiaan Huygens put into practice in 1656. Galileo died at Arcetri on 8 January 1642. His body was privately deposited in the church of Santa Croce; the church forbade any honours to a man who had died under vehement suspicion of heresy. A monumental tomb was built for Galileo 95 years after his death when another Florentine, Pope Clement XII, was enthroned in Rome. This tomb is near the entrance of the church of Santa Croce.

In 1992 the Vatican concluded a 13-year investigation into the 'Galileo affair' by proclaiming that the church had erred in condemning Galileo. The Vatican claimed that this was a result of 'tragic mutual incomprehension'.

Early in his scientific career Galileo had accepted Copernican astronomy but he ignored Kepler's laws of planetary motion, which were discovered during his lifetime. Kepler's purple prose may have been partly responsible. Galileo firmly believed that the orbits of the planets had to be circular in order to keep the fabric of the cosmos in perfect order. This preconception prevented him from giving a full formulation of the law of inertia which he had discovered. The idea of universal gravitation seems to have hovered on the edge of the great man's mind, but he refused to develop it further, believing it (with Descartes) to be 'occult'. Lacking the theory of gravitation, Galileo believed that the inertial path of a body on the Earth must be circular. To Galileo the elliptical orbits of planets proposed by Kepler were simply impossible, and he ignored Kepler's work.

Galileo's greatest contribution was the establishment of mechanics as a science. Some basic facts and theorems had previously been discovered and proved, but it was Galileo who first clearly grasped the notion of force as a mechanical agent. Although he did not formulate the interdependence of motion and force into scientific laws, his writings on dynamics are everywhere suggestive of these laws. He paved the way for Isaac Newton.

4 Newton

By the middle of the seventeenth century the Copernican revolution had brought about a fundamental change in the attitude to nature. The Aristotelian universe of matter with mysterious 'qualities' which gave objects desires and tendencies was under sustained attack. Unsettling scientific views were increasingly gaining a hold on the human mind and the firm association between religious belief, moral principles and the traditional scheme of nature was shaken. It had become increasingly popular to ask 'how' things happened and to demand and provide mathematical exposition and experimental confirmation. Two factors principally encouraged this change: the formation of scientific academies and the development of scientific instruments. In the seventeenth century, the tide of Copernican revolution had flowed past the universities of Europe which were dominated by the Church and did not provide the freedom of inquiry that is taken for granted in universities today. So it fell, instead, to the scientific academies to provide the needed encouragement, support and a forum for communication, essential for dissemination of new results and ideas. The first organisation that could be considered a scientific academy was the *Accademia dei Lincei* (Academy of the Lynx-eyed, the lynx symbolising the sharp eye of science). The society was founded in Rome in 1603 by Duke Federigo Cesi who combined his wealth and curiosity to set up a forum independent of ecclesiastical and university control or prejudice. The Academy was international from its beginning – one of its first charter members being Dutch. The academy members could add the title 'Lyncean' after their name on any literary work they published. This society had frequent meetings at which the members discussed the results of their individual experiments. The society dispersed after Cesi's death in 1630. Some two decades later two Medici brothers, Grand Dukes Ferdinand II and Leopold, founded the *Academia del Cimento* (Academy of Experiments) in Florence. They supplied the academy with a laboratory containing the finest instruments then available in Europe. This society was formally organised in 1657, but its members had met informally since 1651. The society was different from the academy in Rome and from most modern scientific societies. It encouraged its members to work together, rather than individually, on the most important problem of the time. In this respect it was like the modern nationally and multi-nationally funded scientific laboratories, where teams of scientists work jointly on major scientific (or industrial) problems. The society sponsored numerous investigations in physics, such as the thermal properties of solids and liquids and the measurement of the speed of sound. Unfortunately the society lasted only a short time and was disbanded in 1667.

In England the Royal Society began life in 1648 as a small informal group which met in the rooms of an Oxford don, John Wilkins of Wadham College. This 'Invisible College' used to meet at irregular intervals to discuss the latest ideas from Europe and from their own colleagues. By 1659 it had acquired a more formal structure and had a regular meeting place at Gresham College in Bishopsgate, London. Three years later the College and a number of smaller academies in England became incorporated and received a Royal Charter from the newly restored King Charles II. The Royal Society for 'The Promotion of Natural Knowledge' was largely composed of Puritan sympathisers and received little more than moral support from the Crown. The members of the Society were called 'Fellows' and the meetings of the Society were organised around the presentation of papers followed by discussions among the fellows. Within a few years of its foundation, the Royal Society became the arbiter of all scientific activity in England and anyone recognised to have any scientific talent was invited to join. The Society also founded a journal – the *Correspondence* – for fellows to communicate their ideas. The *Correspondence* evolved into the *Philosophical Transactions* by 1665. This was the first journal which contained original communications, including papers read before the Society, and is one that the Society continues to publish today. The remit of the Society was, according to one of its founders, Joseph Glanvill, 'to enlarge knowledge by observation and experiment', but during its early days the Society was very 'open-minded' and considered numerous crackpot ideas along with ideas of real scientific value. The majority of natural philosophers of the seventeenth century had not realised or accepted that if a series of experiments supported an 'idea' then a general law could be established by inductive reasoning. To a modern scientist this is self-evident, there is no other way of 'doing' science – this is the modern scientific method. Isaac Newton was the first natural philosopher to apply this method fully. His 'Theory of Light and Colour', published in 1672, was the first paper in the *Transactions* that presented experimental results to refute an accepted scientific theory.

Like the Royal Society, the French *Académie des Sciences* developed from an informal gathering in Paris of a group of interested natural philosophers and mathematicians. Louis XIV formally established it as a Regular Academy in 1666. Unlike the Royal Society, in France the Crown provided the members of the *Académie* with pensions and financial support for their research. But this financial security was gained at the expense of academic and organisational independence. In the *Académie* research was conducted jointly in the manner of the *Academia del Cimento*. The *Académie* sponsored investigation on a variety of subjects covered by its two sections: mathematics (which included geometry, astronomy and mechanics) and physical sciences (which included chemistry, anatomy and botany). It was reorganised in 1699 and began regular publication of original papers in its *Mémoires*. Following the French revolution, during which all academies were abolished, it was refounded in 1795.

By the start of the eighteenth century almost all European countries had scientific societies.

The scientific societies of the seventeenth century provided relative doctrinal freedom on scientific questions but rigorous evaluation of research by a peer body. Such evaluation is an important requirement of modern science. The separation of research from teaching was perhaps the most striking characteristic that distinguished the academies from the university-based research that developed in the nineteenth century.

The increasingly searching inquiries of the scientists of the late sixteenth and seventeenth centuries required new instruments with which to conduct their experiments. The idea of controlled experiments and the extrapolation of the results to ideal situations were new to scientific inquiry. This need for experimentation brought about the marriage of scientist and craftsman. In the previous century, Tycho Brahe had recognised the value of finely crafted instruments for accurate measurements. In the following decades there was rapid development in precision scientific instrumentation. Galileo's amazing observations of the Jovian moons would not have been possible without the craftsmen who made the glass and polished the lenses he used in his telescopes. The leading scientists of the seventeenth century, Galileo, Hooke and Newton, were accomplished craftsmen. With the precision instruments they built, natural phenomena could be studied under controlled conditions and the results were reliable, and more importantly, repeatable. In the latter half of the seventeenth century there were telescopes, microscopes, barometers, pendulum clocks, vacuum pumps and bubble levels. These instruments, crude as they were by present standards, had a profound influence on seventeenth-century science.

The new natural philosophy that originated in Renaissance Italy gradually spread north and west. In England the process that Galileo had begun, 'to bring the universe down to Earth for experimental examination', was taken to its logical conclusion by one man, Isaac Newton. Newton was the culminating figure of the scientific revolution of the seventeenth century.

ORIGINS OF DYNAMICS

Isaac Newton (1643–1717) was born in Woolsthorpe, Lincolnshire. It is generally believed that Newton was born in 1642, the year Galileo died. This confusion is a result of the different calendars in use in England and Continental Europe in the later half of the seventeenth century. Italy had adopted the revised Gregorian calendar but the old Julian calendar was still in use in England. By the Julian calendar Newton was born on 25 December 1642 which by the Gregorian or the modern calendar is 4 January 1643; thus Newton was born almost a year after Galileo's death.

Not much is known of Newton's childhood; his father had died before his birth and he was a frail child and remained a hypochondriac all his life. When he was three years old his mother remarried and went to live with her new husband, and young Newton was brought up by his maternal grandmother. At 12 he went to King's School in Grantham, about seven miles from Woolsthorpe, where he lodged with the family of the local apothecary. The formal education at King's and the encouragement from the apothecary to 'play' with chemicals appear to have laid the foundation of his intellectual development. In 1661, against his mother's wishes, Newton left for Cambridge where he spent the next 30 years of his life.

At Cambridge Newton enrolled at Trinity College on 5 June 1661 and for a time seems to have been a conscientious and dedicated student but not exceptional in any way. He was also a reclusive young man who found it difficult to make friends, mix with other students or take part in the general undergraduate activities. Aristotelianism formed the core of higher education in European universities when Newton arrived at Cambridge and this would have been the focus of his early studies. Newton was exposed to the latest scientific developments and writings of Copernicus, Tycho Brahe, Kepler, Descartes, Galileo and others (except Galileo's *Dialogue Concerning the Two Chief World Systems* and *Dialogue Concerning Two New Sciences*, both of which appear to have been too risqué for the Cambridge authorities). Newton mastered Descartes, who viewed physical reality as composed entirely of particles in motion and natural phenomena as a result of this motion. He was also exposed to the works of Cambridge Platonists and through them was introduced to the magical Hermetic tradition, which sought to explain natural phenomena in terms of alchemical and magical concepts. These two opposite traditions of natural philosophy continued to influence Newton's thoughts for the rest of his life. During his second year at Cambridge, Newton appears to have undergone a radical change of thought and he stepped away from the traditional scholastic approach to natural philosophy (science) and began to question what he was taught. He convinced himself that scientists could not simply trust what they observed with their senses, but must perform experiments in order to understand the nature of the universe. In 1664 Newton bought a prism at a country fair and started a programme of experiments to investigate the nature of light. These experiments appear to have been attempts to reproduce the results described in Descartes' book on colours, and he would later describe them in *Opticks*, published in 1704. About this time Newton also decided to study mathematics and it is possible that he attended the series of lectures given by Isaac Barrow, the first Lucasian Professor of Mathematics at Cambridge. But this early education in mathematics was unguided and when Barrow examined him for his undergraduate scholarship he was found not to know Euclidean theorems. Barrow probably did not think much of Newton as a mathematician, but he passed him nevertheless for his scholarship. Newton, made aware of this deficiency, immediately undertook a rigorous study of Euclidean geometry and algebra. In the spring of 1665 Newton graduated, with a second-class BA. Graduation secured his future at the university and Newton decided to devote himself to research to unravel the laws of nature.

The plague of 1665 played a crucial part in Newton's life. The plague started in London and gradually spread through England. Newton left Cambridge in the summer of 1665 (the college was dismissed on 8 August) and travelled north to Woolsthorpe, to his mother's home, not returning to full time residence in Cambridge until 1667. During his absence from Cambridge Newton laid the foundation of calculus and extended his studies of optics. And, according to the popular tale, the impeccably timed fall of the second most famous apple in (Christian) history, in the orchard next to his mother's house, set in motion the great synthesis of the motion of celestial bodies and the motion of bodies on Earth, or the formulation of the theory of universal gravitation. The story was popularised by the French writer Voltaire, Newton's great admirer, and was helped along by Newton himself. Whatever the truth of the story, there is no doubt that the inspiration which led to the discovery of the laws of motion and universal gravitation was provided during his stay at Woolsthorpe. However, it would be misleading to believe that Newton had this inspiration out of the blue. The 'laws of motion' formed a 'hot' topic of research and Descartes, Galileo, Huygens and others had considered it before Newton.

Descartes had published some of his thoughts on motion in *Principia Philosophicae* in 1644, but, aware of the trials of Galileo, left others in manuscript form. This was no loss to science as a number of his 'thoughts' were speculative and not real observations or experiments. Some of these manuscripts were published posthumously in his *Le Monde* in 1664. The following three concepts (described in Chapter 3, Galileo) summarise his thinking behind his 'laws' of motion: (1) he believed space and time to be relative; (2) he offered erroneous laws of collision, and (3) he advocated the vortex model of planetary motion. Descartes rejected the notion that bodies could interact through empty space. He insisted that force must be propagated by a material substance (called ether) and therefore all space must be filled with this substance.

Christiaan Huygens (1629–1695) developed Descartes' ideas further. Huygens was from a wealthy and distinguished middle-class Dutch family. Very early he showed a marked mechanical ability and a talent for drawing and mathematics. Some of his early geometrical work impressed Descartes, who was an occasional visitor at the Huygens' house. In 1645 Huygens entered the University of Leiden, where he studied law and mathematics. Ten years later he visited Paris for the first time, where his distinguished parentage, wealth and affable personality gave him entry to the Parisian intellectual and social circle. Back in the Hague he invented new methods for grinding and polishing lenses thus improving the performance of telescopes. With these improved telescopes he discovered a satellite of Saturn in March 1655 and a year later distinguished the stellar components of the Orion nebula. In 1659 he also discovered the true shape of the rings of Saturn, which Galileo had missed. In 1666 he became one of the founding members of French *Académie des Sciences* and lived in Paris for the following 15 years.

Huygens rejected certain Cartesian tenets but he always reaffirmed his belief that mechanical explanations were essential in science. This was to influence his mathematical interpretation of both light and gravitation. In 1673 he published his masterly *Horologium Oscillatorium*, a work which was to have a profound influence on physics through its exposition of the principles of dynamics and complete derivation of the formulae for the time of oscillation of a simple pendulum, the oscillation of a body about a stationary axis, and the laws of centrifugal force (an unfortunate name as this is not really a 'force' but a reaction against a restraining action) in uniform circular motion. The book gives a mathematical analysis of the pendulum clock and devotes attention to the composition of forces acting in circular motions. In this work Huygens gives a clear exposition of the nature of inertia. He wrote:

> If gravity did not exist nor the atmosphere obstruct the motion of bodies, a body would maintain for ever, with equable velocity in a straight line, the motion once impressed upon it.

Like Descartes before him, Huygens delayed publication of his work on mechanics but this was not through any concern for ecclesiastic displeasure; he was unhappy with the presentation of his works. This delay in publication was singularly unfortunate, as Newton had to rediscover a number of steps Huygens had already worked through.

In 1681 Huygens returned to Holland (prompted by serious illness) and stayed there for the rest of his life because return to Paris was made impossible by the increasingly reactionary (anti-Protestant) attitude of Louis XIV. He visited London in 1689, met Newton, and lectured at the Royal Society on his own theory of gravitation. This was published in 1690 in his *Discourse de la cause de la pesanteur* (Discourse on the cause of Gravity). Huygens' theory of gravity was based on Cartesian vortices. Although Huygens did not enter into public controversy with Newton, it is evident from his correspondence, particularly with Leibniz, that he regarded any theory of gravitation that was devoid of a mechanical explanation as fundamentally flawed. Huygens was a formidable mathematician; he acknowledged that Newton's assumption of forces acting between planets of the solar system was justified by the correct conclusions that could be derived from it. But he could not accept the notion of 'action at a distance' required in Newton's theory. This he felt was dangerously close to 'occult'. Huygens and also Leibniz had misconstrued the 'occult' nature of Newton's 'action at a distance'; Newton was very clear that this was a 'working hypothesis' and not a physical mechanism (see later). Huygens' theory of gravitation was never taken seriously and remains today of historical interest only. But his theory of rotating bodies and his contribution to the theory of light were of lasting importance.

Early in 1667 Newton returned to Cambridge, as the university had reopened, and started working for his fellowship, which he was awarded later that year. A year later he obtained his MA and rose to the rank of major fellow. The fellowship brought Newton a small stipend and a job for life. Around this time Isaac Barrow became convinced that during his two years away from Cambridge Newton had made discoveries of profound significance. Barrow also realised that Newton was extremely secretive and had deep-rooted suspicions that others would steal his discoveries. He was, therefore, very reluctant to publish his work. In 1669 Newton wrote his work on mathematics in *De Analysi per Aequationes Numeri Terminorum Infinitas* (On Analysis by Infinite Series). This was circulated in manuscript to a limited circle of academics and it brought Newton recognition among his peers. Two years later this paper was revised as *De methodis serierum et fluxionum* (On the methods of series and fluxions). The word 'fluxions' was invented by Newton and heralded the birth of calculus. Unusually sensitive to questions of rigour, Newton based his calculus on a sound foundation of ideas from kinematics. A variable was known as a 'fluent', a parameter that changes with time, its derivative or rate of change with time was called a 'fluxion'. Calculus allowed a mathematical description of continuous change – given a starting configuration, past and future configurations could be reconstructed. Newton had originally developed calculus to analyse planetary dynamics because the force on a planet in orbit changes continuously. Barrow was sufficiently impressed to recommend Newton for the Lucasian Chair and in 1669, aged 26 years, Isaac Newton became the second Lucasian Professor of Mathematics at Cambridge – a position he continued to hold for several years, even after leaving Cambridge in 1696. Three years later, in 1672, he was elected a Fellow of the Royal Society for his invention of the reflecting telescope.

During the next 20 years Newton continued to develop the conceptual foundation of the theory of universal gravitation which he had laid during the two 'plague' years he spent in Woolsthorpe. He developed a detailed theory based on both mathematically rigorous analysis and experimental verification. Both these elements were necessary. Without mathematics his intuitive insight into the universality of gravitational attraction and its action at a distance would have remained just 'another good idea'. Without the experimental verification his theories would not have achieved the status they did. At Woolsthorpe Newton had applied the inverse square law of gravitation to circular motion; by 1680 he was able to show that the same law of gravity also accounts for the elliptical orbits of planets with the attracting body located at one focus of the ellipse.

In November 1680 there appeared in the sky what was originally thought to be the first of two comets. This comet was observed by astronomers all over Europe. Newton also made detailed observations and continued these for about a month till the comet seemed to disappear. Newton, like all other astronomers, had not realised that the comet's motion was influenced by the Sun, and he might never have realised this had not a 'second' comet appeared in the sky in autumn 1682. This comet was also followed by a large number of astronomers in Europe, and its motion was soon realised to be retrograde, that is, it was moving away from the Sun – it was the 'first' comet, which had travelled round the Sun and was moving towards the outer reaches of the solar system. Between the sightings of 1680 and 1682 Newton had started calculations to determine the orbits of these comets using the methods he had developed in 1666 and had applied recently to the elliptical paths of planets. He calculated a number of orbits, eventually settling on an elliptical orbit with gravitational force acting between the centre of the comet and the centre of the Sun. He concluded that the motion of the comets obeyed the same inverse square law that he had demonstrated to apply to planets and the Moon. However, Newton was uneasy with force acting at a distance without an observable mediating medium or mechanism. Descartes' mechanical philosophy still had a strong hold on his thinking, and this described gravity as arising from vortices within the ether which was visualised as a weightless invisible medium pervading all space and facilitating all action. Newton had rejected almost all aspects of Descartes' mechanical theory, including his theory of origin of gravity. However, he was not yet ready to abandon the ether – the mediating medium for a force to act at a distance. In the early 1680s, he decided to calculate the velocity of planets in their orbits, to check the validity of Kepler's Third Law in the context of his theory of gravity. He now had calculus to make these calculations. The results stunned him: the calculated paths of the planets (made assuming movements through vacuum) matched the observed paths precisely. Only one conclusion was possible – there was no ether through which the planets moved. If there were such a medium then it would slow down the planets in their orbits. Newton decided to test this in the laboratory by observing the motion of a pendulum. A swinging pendulum slows down, the slowing being due to friction at the point from which the bob is suspended, the resistance of air and the possible resistance of the ether. The presence of ether could be deduced by observing the period of a pendulum with a (heavy) hollowed bob – first with the bob empty and then with the bob filled with various amounts of different materials, to see if the amount and the type of material in the pendulum affected the retardation of the pendulum. He determined that the rate of retardation was independent of the amount or the type of material. This convinced him that the pendulum did not move through physical ether or the ether did not interact with the material of the pendulum. Newton concluded that if ether existed at all, it was almost a vacuum. With these experiments Newton was ready to abandon *all* aspects of Descartes' mechanical universe and embrace the concept of 'action at a distance'.

Ether, it turned out, was 'down but not dead'. In the early nineteenth century Newton's corpuscular theory of light gave way to a wave theory of light and the concept of an all pervading universal ether was revived to explain the propagation of light waves. The universal ether was only abandoned after very precise optical experiments (see Chapter 5, Einstein).

By the autumn of 1684 all the elements required for a fundamental revision of the laws of dynamics were in place. But Newton did not live in a culture of 'publish or perish', as modern scientists do, and the publication of his work on mechanics and gravitation had to await the intervention of Halley. Edmond Halley (1656–1742) was the second Astronomer Royal, succeeding Flamsteed at Greenwich in 1720. He was an astronomer of remarkable abilities. Before he was 20 he had determined discrepancies between the theoretical and observed paths of Jupiter and Saturn. Influenced by Flamsteed's programme to determine accurately the positions of northern stars, Halley proposed to do the same for stars visible from the Southern Hemisphere. In 1676 he sailed south (without taking his degree at Oxford) to St Helena in a ship of the East India Company. Bad weather (the bane of all ground-based astronomy) prevented satisfactory completion of this programme. But in January 1678, he returned to England with measurements of the longitude and latitude of 341 southern stars, and observations of the transit of Mercury across the disc of Sun. His star catalogue was published in 1678; this was the first star catalogue of southern stars made with positions determined telescopically. It established Halley's reputation as an astronomer, and he was elected to the Royal Society.

In 1680 Halley started a study of the orbits of comets and was lucky enough to study in detail the path of the comet which appeared in 1680 and 1682. Halley calculated the orbit of this comet and found that it was very similar to the orbit of comets of 1531 and 1607. He surmised that he was observing the return of the same comet. In 1705 he was to conclude that this comet returns every 75.5 years, following a very elongated elliptical orbit extending beyond the orbit of the planets. Halley, with Robert Hooke and Christopher Wren (the famous architect), was attempting to determine the mechanical force that kept comets and planets in orbits around the Sun. Both Halley and Hooke had calculated that a force that decreased as the square of the distance kept the planets in orbit but they were not able to determine from this hypothesis the theoretical orbits of the planets. Prompted by his growing interest in the nature of planetary motion in 1684, Halley travelled to Cambridge to discuss the problem with Newton. He was aware of the Lucasian Professor's interest in mechanics, optics and gravity. Halley wanted to know 'the curve that would be described by the planets supposing the force of attraction towards the Sun was the reciprocal of the square of their distance from it'. Newton replied immediately that it would be an ellipse. Halley asked him for a mathematical verification, but Newton had mislaid his proof and Halley returned to London disappointed. But three months later Newton sent him a nine-page paper entitled *De Motu Corporum in Gyrum* (On the Motion of Revolving Bodies). This paper was presented to the Royal Society on 10 December 1684. In many ways *De Motu* is a very surprising paper. In this paper Newton demonstrates a dynamics that leads to Kepler's three laws. But the three laws of motion with which Newton's name is associated are not mentioned explicitly. These came later when Newton set about revising and expanding *De Motu*. This suggests that in 1684 Newton had not yet arrived at a consistent and rigorous description of dynamics.

Halley continued his research making practical application of his work by compiling meteorological and magnetic charts of the Atlantic and Pacific. In 1705 he published his *A Synopsis of the Astronomy of Comets* in which he described the orbits of 24 comets that had been observed from 1337 to 1698. He showed that the orbits of the three historic comets of 1531, 1607 and 1682 were so similar that they must be the successive returns of the same comet, and he predicted that the comet would return in 1758. This comet is now known as Halley's comet. In 1716 he devised a method for observing the transit of Venus across the disk of the Sun, predicted for 1761 and 1769. Such observations could be used to determine accurately, by parallax, the distance to the Sun. In 1720 Halley was appointed the Astronomer Royal at Greenwich.

Encouraged by Halley (and goaded by an acrimonious exchange of letters with Hooke regarding the path of an object falling down to the revolving Earth) Newton started work on a comprehensive exposition of the laws of motion and the laws of universal gravitation. This great work appeared in 1687 under the title *Philosophiae Naturalis Principia Mathematica* (Mathematical Principles of Natural Philosophy) and is better known today by the abbreviated title, the *Principia*. The *Principia* was edited and published by Halley 'at his own charge' and consists of three separate but linked books and an introduction. The celebrated three laws of motion are stated in the introduction and Book I. A major part of Book I is based on *De Motu* in which Newton explains the concepts of centripetal force. Book II deals with motion through a resisting medium and the motion of such a medium. Here Newton explores the dynamics of vortices. In the conclusion to this Book Newton demonstrates that Descartes' theory of planetary vortices cannot be true. In Book III Newton introduces his theory of gravitation. He goes on to describe the unification of the terrestrial mechanics of Galileo (which was not expected to be related in any way to the motion of heavenly bodies) and Kepler's laws of motion of planets (which were not expected to have any relationship with earthly movements). This was Newton's great accomplishment – the concept of universal gravitation.

Newton's three laws of motion are the principles of dynamics, which replace the 'natural motions' and the 'violent motions' of Aristotelian mechanics.

Newton was not the first to arrive at the concept of gravitation. As early as the fifteenth century some astronomers had considered a force of attraction between the heavenly bodies and the Earth. Gilbert in 1600, Bouillard in his book *Astronomica Philolaica* published in 1645 and Borelli came close to the basic features of the law of gravitation. Kepler actually considered an inverse square law before rejecting it. But it was Newton who brought the different strands together into the theory of universal gravitation.

LAWS OF MOTION

Newton wrote the *Principia* in classical Latin and suppressed its publication in English until 1727, the final year of his life. It is not a book for the uninitiated, being written as a series of propositions, each of which has to be fully understood before the following one can be tackled. Moreover, Newton does not give the details of his calculations, which makes it difficult to follow the mathematical expositions. Surprisingly (or deliberately) Newton rejected algebraic methods (so important in his research on calculus) in the *Principia*, the mathematical basis of his dynamics was the geometric theory of limits. Only a few hundred copies of the first edition of the 550-page *Principia* were sold. But it has since gone through almost 100 editions. A detailed exposition of the parts of the *Principia* that deal with the laws of motion and the formulation of the universal law of gravitation has been published by the late Indian-American physicist Subrahmanyan Chandrasekhar, in his *Newton's Principia for the Common Reader*. Extracts quoted from the *Principia* have been taken from this book.

In the early months of 1685 when Newton began to revise and expand *De Motu* he realised that a consistent theory of dynamics demanded a rigorous definition of concepts which were pertinent to his theory. These definitions and the laws of motion precede the formal opening of Book I.

> *Definition I.* The quantity of matter is the measure of the same arising from its density and bulk conjointly

This introduces the concept of mass of a body, which is defined as the product of density and volume of the body. The density is not further defined which suggests that Newton regarded density to be a fundamental quantity instead of mass. There is a further level of imprecision – matter is not defined. Later Newton equated mass with inertia, which is the modern view. To define mass without reference to inertia is incorrect. The definition of (and the name) mass and its central position in a consistent development of dynamics was a major discovery by Newton.

> *Definition II:* The quantity of motion is the measure of the same, arising from the velocity and quantity of matter conjointly.

Nowadays the quantity of motion is defined as momentum and is the product of mass and velocity.

> *Definition III:* The *vis insita,* or innate force of matter, is a power of resistance, by which every body, as much as in it lies, continues in its present state, whether it be of rest, or of moving uniformly forward in a right line.

This is the definition of inertia. In *De Motu* Newton had not arrived at the principle of inertia: he still believed that a body in uniform motion was carried along by a force inherent in it. During 1685 as he worked on and developed *De Motu* into a general science of dynamics that would describe equally well all motion – free fall on the surface of Earth and orbital motion in the heavens – he realised the centrality of the principle of inertia. He thus arrived at the definition given above and the word *inertia* entered the language of science. The definition of mass as the 'quantity of matter' in a body and that of inertia as the 'property of a body to resist any change in its motion' are both rather pedestrian and tell us very little about the nature of mass or inertia. These concepts have actually proved remarkably difficult to pin down and unambiguous ideas about the 'origin of mass' only began to emerge in the last quarter of the twentieth century (see Chapter 8, Planck).

> *Definition IV*: An impressed force is an action exerted upon a body, in order to change its state, either of rest, or of moving uniformly forward in a right line.

This is the definition of force. In tribute to Newton the modern unit of force is a 'newton' (in the International System of units). A newton is the force needed to accelerate a body weighing one kilogram by one metre per second per second.

> *Definition V*: A centripetal force is that which bodies are drawn or impelled, or any way tend, towards a point as to a centre.

The concept of centripetal force (*vis centripeta* – seeking the centre) was central to *De Motu* and was introduced by Newton to replace the then current and confusing centrifugal force (fleeing the centre). In this definition Newton appears to have intended to include all action-at-a-distance forces. At this stage he had not identified gravity on Earth with the force holding the planets in orbits although he seems to have suspected this. He states that it is a force, whatever it may be, which draws a planet from the rectilinear motion that it would pursue, and makes it revolve in a curvilinear orbit.

After several further definitions pertaining to centripetal force, and explanatory notes (*scholium*) on relative and absolute quantities, Newton gives his three laws of motion. These laws are the starting points of every argument in classical dynamics. The first two laws, which relate to inertia of a body, were generalisations from Galileo's experiments and observations. The three laws of motion in the introduction to the *Principia* are:

> *First Law*: Every body continues in its state of rest, or of uniform motion in a right line, unless it is compelled to change that state by force impressed upon it.

This is known as the law of inertia and is a special case of the Second Law given below. This law describes the motion of a body in equilibrium, that is, the motion of a body when no force is acting on it.

> *Second Law.* The change of motion is proportional to the motive force impressed; and is made in the direction of the right line in which that force is impressed.

By change of motion Newton had in mind the rate of change of momentum. For the case of constant mass, this definition becomes *force = mass × acceleration* (the second law was put in this form, which is the modern form, by the Swiss mathematician Leonhard Euler in 1750). Thus the Second Law provides a definition of force – acceleration given to a mass. This law describes the motion of a body when there is a force acting on it which is not balanced by another force.

> *Third Law.* To every action there is always opposed an equal reaction; or, the mutual actions of two bodies upon each other are always equal, and directed to contrary parts.

This law was original to Newton. Together with the Second Law this law describes the concept of mass in terms of the inertia of a body – inertia is the power to resist a change of state and this can be parameterised by mass. Mass can only be described in terms of inertial or gravitational properties of a body and not independently of the concept of force. The first two laws describe the motion of an 'isolated' body and say nothing about the source of the force causing the motion nor anything about the relationship between the source of the force and the action caused by the force. The Third Law achieves this – this law states that the body exerts an opposite force on the source of the force and the two opposite forces are equal in magnitude. It should be emphasised that to set a body in motion the two opposing forces do not act on the same body, if this were true then the body would not be accelerated. The opposite forces act on different bodies or media. For example, consider a boat being rowed: the pushing force moves the boat forward but the equal and opposite force acts against the water or the resistance of the water.

The study of motion necessarily involves the concepts of *space* and *time*. Newton formulated his three laws to describe the motion of bodies, but motion relative to what? The ideas that motion must be *relative*, that is, displacement with respect to some reference system, was first identified by Galileo (but had also been assumed by Copernicus and Kepler). Passengers in a totally enclosed car, which is moving at a uniform speed, will not realise that they are in motion. They will only realise their motion if they can see the stationary buildings or trees around them. Newton believed that everything moved in relation to a fixed but undetectable spatial frame so that it could be said to have an absolute velocity. Time also flowed at the same steady pace everywhere. Even if there were no matter in the universe, the frame of the universe would exist, and time would flow even though there was no one to observe its passage. The principle of inertia asserts that there exists a class of frames of reference that are neither accelerated nor rotated. These are called inertial frames of reference. Bodies not subject to external force move with respect to inertial frames of reference at constant speed in a straight line. This implies that it is impossible, by purely mechanical means, to distinguish one inertial frame of reference from any other inertial frame. Newton's laws of motion are invariant (they do not change) when transformed between two frames in uniform relative motion. The equivalence of inertial frames is now known as the principle of relativity (the phrase 'the principle of relativity' was not used in the seventeenth century but was introduced by the nineteenth-century French physicist Henri Poincaré). Newton had encountered relative space and time in the writings of Descartes and he believed this to be dangerously close to atheism. He therefore introduced the concepts of absolute space and time. In the *Principia* Newton defined *absolute motion* as 'translation of a body from one absolute place to another', where an absolute place is an absolute position in space. Newton does not give the meaning of 'absolute position' – that is left as an intuitive concept. He similarly described *absolute time* thus: 'absolute, true and mathematical time, of itself, and from its own nature flows equally without regard to anything external'.

He also regarded time and space to be independent. That is, all events have a distinct and definite position in space and occur at a particular moment in time. If two events are observed to take place simultaneously by an observer in one inertial frame, they will appear simultaneous to all other inertial observers. That is, everything from the fall of an apple on Earth to an exploding supernova in a distant galaxy is connected by a single coordinate system with just one moment of 'now', or one origin. To put it another way, once synchronised, identical clocks keep time with one another regardless of their state of motion. If this was true then Newton's mechanics would have fantastic descriptive power – any one event in a closed system (that is a system with no external forces acting on it) would provide information about *all* the past or the future events in the system. Newtonian space and time are also homogeneous and isotropic: that is, the properties of time and space are same everywhere and at all times and contain no singularities. Newton defined 'distance' as the spatial separation in space between two points and this separation between two simultaneous points does not alter in space or time (it is invariant). To Newton the invariant distance was self-evident; it had been recognised centuries before Newton that the distance between two points is same irrespective of how and when it is measured. Invariant distance is a characteristic of Euclidean geometry and Newton had assumed (although he does not mention this in the *Principia*) that the geometry of space is a flat, three-dimensional continuum, so that Euclidean geometry applies to all possible arrangements of point locations. To put it other way, in Newton's universe the sum of angles of a triangle is 180 degrees everywhere and at all times.

Newtonian dynamics was, however, formulated for inertial systems, and the questions of absolute space and time are purely academic and do not affect the dynamical laws that he formulated. The problem of relative and absolute motion becomes relevant only when the motion of the frame of reference affects the dynamical laws, as happens in electrodynamics. The universality of time and space is so deeply ingrained in the human mind that it is almost impossible to conceive of alternatives. At the start of the twentieth century the absolute character of space and time were called into question. Albert Einstein, in formulating his

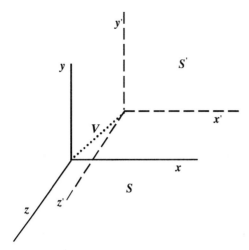

FIGURE 4.1 Two frames in uniform relative motion with speed V. The positions of arbitrary points S and S' in either frame are related by Galilean transformations.

theory of general relativity, abandoned these concepts along with invariant distance and Euclidean geometry and only retained the principle of relativity (more about this in the next chapter).

A frame of reference can be thought of as a Cartesian coordinate system to define the spatial positions of events along with a clock to describe the times of occurrence of these events. The motion of a body in a frame of reference can then be specified by giving its spatial position as a function of time. Consider two frames S and S' (Figure 4.1) in uniform relative motion with velocity V. The coordinates of the position of a body in the frame S' are related to the coordinates in frame S as follows:

$$x' = x - V_x t$$
$$y' = y - V_y t$$
$$z' = z - V_z t$$
$$t' = t$$

where the undashed quantities are parameters in frame S, the dashed quantities are parameters in frame S', and V_x, V_y and V_z are components of the relative velocity of the two frames along the three Cartesian coordinates. This is known as a *Galilean transformation*, after Galileo.

In this transformation the time coordinate is identical and equal to an 'absolute time' in both frames.

UNIVERSAL GRAVITATION

It is possible that Newton was inspired by a falling apple to think of gravity and to discover that the force that pulls an apple to the ground also keeps the planets and the Moon in their orbits. In Books I and II Newton laid down the basic principles of dynamics. In Book III he applies these principles to the solar system to demonstrate that they imply the existence of a cosmic attractive force. This follows from his First Law of motion: because the planets are moving along curved paths (elliptical or circular, it does not really matter) and not in straight lines, there must be a force acting on them. Newton identified this force with the cause of heaviness (*gravitas*) on the surface of the Earth. In the *Principia* Newton treats celestial motions as problems in mechanics and shows that the same principles of motion that account for phenomena on the surface of the Earth also account for all the phenomena in the heavens. This was Newton's great synthesis, that of Kepler's three laws of planetary motion and Galileo's kinematics of uniformly accelerated motion. This synthesis emerged as a natural consequence of the set of dynamical principles that Newton had formulated. That the same set of laws describes the motion of falling bodies and the motion of planets is not obvious and Newton's great genius was to make this counter-intuitive link. Newton thus demolished Aristotle's contention that celestial and terrestrial bodies were subject to different laws. In Book III Newton concluded that all bodies attract each other. To ensure that there was no misunderstanding about the forces between planets, he states:

> The force which retains the celestial bodies in their orbits has been hitherto called centripetal force; but it being now made plain that it can be no other than a gravitating force, we shall hereafter call it gravity. For the cause of that centripetal force which retains the Moon in its orbit will extend itself to all the planets.

Newton used Kepler's Third Law to demonstrate that the force of gravity is inversely proportional to the square of the distance from the Sun (this is the *inverse-square law of gravity*). He proved mathematically that the gravitational attraction of spherical bodies of uniform density could be assumed to act from the centre of the sphere with the total mass of the sphere concentrated at the centre. The gravitational attraction of a non-spherical body can be assumed to act from the *centre of gravity* of the body. Because of symmetry, the centre of gravity of a spherical body is at its centre. Also because of the symmetry of the spherical body and the inverse square law of gravitation, *within a spherical body of uniform density* the net gravitational attraction is zero. Newton's law of universal gravitation is:

> That there is a power of gravity pertaining to all bodies, proportional to the several quantities of matter, which they contain.

The magnitude of this force is proportional to the product of the mass of the two bodies and inversely proportional to the square of the distance between them. Mathematically this can be represented as:

$$F = G \frac{m_1 \times m_2}{r^2}$$

where m_1 and m_2 are the masses of the two mutually attracting bodies, r is the distance separating them and G is the gravitational constant (this constant does not occur in the *Principia* but was introduced by Laplace in the eighteenth century). This statement of universal gravitation is remarkably simple and symmetric. The force depends only on the distance separating the two masses, and not on the orientation, velocity or acceleration of the masses. In formulating the law of universal gravitation, Newton had assumed implicitly that gravity acts at a distance by an as yet unknown mechanism and that it acts instantaneously. The gravitational constant is *assumed* to be positive, as gravitational force is only attractive, repulsive gravitational force has never been encountered.

The mass of a body appears in two laws of Newton: the law of motion, where the mass of the body is the inertial mass (that is, inability to change a state of rest or motion), and the law of gravity, where the mass of the body is the gravitational mass or weight. Weight is defined as the force with which a body is attracted gravitationally by the Earth (this force can be measured by a spring balance). The weight of a body can change if it is taken to another planet or even to different locations on Earth, because of the flattening of the Earth at the poles. For example, a body that weighs *1 kg* at the equator will weigh *0.17 kg* on the Moon but *2.64 kg* on Jupiter. The same body will weigh *0.25* per cent more at the poles of the Earth. But the mass of the body is the same at all these locations. The precise distinction between 'mass' and 'weight' is not possible without reference to the Second Law of motion. Newton noted that the (inertial) mass 'is proportional to weight, as I have found by experiments with pendulums'. The experiment consisted of two pendulums, each 11 foot (about 3.35 metres) long. The bobs of both pendulums were identical (to ensure that the resistance of air was similar for the two pendulums). One pendulum was a reference pendulum and in the bob of the other he put gold, silver, lead, glass, salt, water, etc. He found that the periods of the two pendulums were independent of the material of the bob and were similar to one part in 1000. Newton was to state:

> That all bodies gravitate towards every planet; and the weights of bodies towards any one planet, at equal distances from the centre of the planets, are proportional to the quantities of matter which they severally contain.

This is now known as the *weak equivalence principle*. Newton then makes the following remarkable statement:

> But, without all doubt, the nature of gravity towards the planets is the same as towards the Earth.

Newton does not give a proof in support of this statement but his reasoning was that the Moon and the satellites of Jupiter obey the same laws of orbits (Kepler's laws) and therefore the Jovian gravity, like Earth's gravity, obeys the same inverse square law. If the satellites were assumed to fall towards Jupiter then they would move equal distances if dropped from equal heights. Similarly if planets are assumed to fall towards the Sun then starting from equal heights they will fall at equal rates. Newton is asserting that '*all* bodies have the same acceleration in free fall or in a gravitational field'. This is now called the *strong equivalence principle*. The universal equality of the inertial and the gravitational mass of a body is at the core of Newton's argument for the universality of his law of gravitation. The proportionality of these two masses was to become the central pillar of the theory of general relativity (see Chapter 5, Einstein, and Chapter 6, Dicke).

NEWTON'S UNIVERSE

In 1690 Newton made a bold attempt to apply his laws of universal gravitation to the largest physical system – the universe. Newton soon realised that gravity leads to a highly unstable universe. Consider a large but *finite* spherical system. Suppose the matter in the sphere is uniformly distributed and is initially at rest. Each part of matter in the sphere will attract all other parts gravitationally and the sphere as a whole will begin to collapse towards the centre – this is called self-gravity. Thus a spherical system filled with a uniform distribution of matter will undergo gravitational collapse and the sphere will shrink. Such a sphere will also attract any matter outside the spherical system. If the system were large but finite and not spherical then differential self-gravity will cause the matter to collapse into a large number of lumps of matter. But if the system were *infinite*, like the universe, then the situation would be very different. A finite uniform sphere collapses towards its centre, but an infinite uniform system has no centre! A typical point in a uniform infinite system is similar to all other points and will be pulled equally in all directions and therefore will not move. Thus a static uniform universe is possible under Newtonian gravitation. But, as Newton realised, a small departure from uniformity anywhere would lead to an imbalance and the universe would break up into finite sized clumps of matter, which would collapse under self-gravity. Newton puts it succinctly (and graphically) thus, in a letter[1] to Richard Bentley dated 17 January 1692/3:

> I agree with you that if matter evenly diffused through a finite space not spherical, should fall into a solid mass, ... And much harder it is to

suppose that all the particles in an infinite space should be so accurately poised one among another as to stand still in a perfect equilibrium. For I reccon this is as hard as to make not one needle only but an infinite number of them (so many as there are particles in an infinite space) stand accurately poised upon their point.

LAWS OF KEPLER AND NEWTON

Both Kepler and Newton formulated laws of nature, but these laws differ in a fundamental sense. Kepler's laws are empirical, derived by trial and error from Tycho Brahe's observations and after long tedious arithmetic. These laws lack the predictive power of laws of physics. They can neither be used to forecast where a planet will be at a future date nor be used to send a space probe to other planets or comets. Empirical laws, like those of Kepler, are valid only within the limits of the accuracy of the data from which they are derived or which they attempt to reproduce. Ptolemy and the Arab astronomers of the twelfth and thirteenth centuries had the mathematical tools and the mathematical ability to deduce that the orbits of the planets were elliptical, but the accuracy of their data was too poor to arrive at this conclusion. They wielded 'Occam's razor' and deduced the simplest shape that fitted their data, and they arrived at circular orbits.

Newton's laws are *assumptions* about the nature of the physical world. These laws can have no validity unless they can be proved by experiments and they can reproduce observations. If they fail then they have to be abandoned – even *one* exception will destroy the hypothesis. These laws are also described in terms of ideal conditions – frictionless surfaces and motion through perfect vacuum with no resistance. Such ideal conditions, of course, are never encountered in real life – there is no way to create force-free environments, a body will feel all sorts of forces all the time. Newton (and Galileo before him) had arrived at his ideal laws through observations of a large number of non-ideal experiments. This level of abstraction is the hallmark of modern science. Newton also introduced 'the general method of proof'. A striking example is his demonstration that the conclusions of terrestrial gravity can be extended to planets. Newton stated this beautifully in Rule II of his 'Rules of Reasoning' which are given at the beginning of Book III:

> Therefore to the same natural effect we must, as far as possible, assign the same cause.

This 'inductive extrapolation' is the principal weapon in the armoury of modern physics and has been used to stretch our imagination to the origins of the universe.

Newton's laws have huge predictive power. They are also general laws in the sense that they can be used to investigate the motion of *any* two or any number of moving or colliding bodies. Kepler's law can only describe the shape and motion of a particular orbit for which observed data are available. Newton's general laws suggest that a whole family of orbits, defined by conic sections, is available to a small body moving under the gravitational influence of a massive body. The ellipse is but one of a number of orbits that the small body can have. A particular orbit is defined entirely by the speed of the small body and this speed is uniquely determined by the mass of the larger body and the separation between the two bodies. For one unique velocity the orbit of the small body will be circular. Any departure from this value (due to perturbations) would make the orbit an ellipse. That is why the orbits of planets and satellites in the solar system are elliptical: their motion is perturbed by the gravitational attraction of other planets of the solar system. But the perturbation by other planets is not very large and the deviation from circular orbits is small. Similarly, the orbit of a body projected towards the Sun (e.g. a comet) will be parabolic if the velocity of the projected body is *exactly* equal to the value given by the mass of the Sun and the separation between the Sun and the projected body. If the velocity is less, the orbit will be elliptical and the body will be captured by the Sun. If the velocity is higher then the orbit will be a hyperbola and the body will escape from the Sun. Periodic comets (like Halley's comet) have elliptical orbits. They probably entered the solar system at high speeds but gravitational perturbations by planets of the solar system reduced their speed sufficiently for the comet to be captured by the Sun. All nonperiodic comets have hyperbolic orbits.

Analytical laws, such as those of Newton, are valid as long as their predictions agree with observations. It was the lack of agreement between the observed orbit of Mercury and the orbit calculated from Newton's laws of gravitation that cast doubt on the generality of Newton's laws of gravitation (see Chapter 5, Einstein).

CONSERVED QUANTITIES

Physicists attempt to describe natural processes as economically and simply as possible. Simple, in this context, means with the most generally applicable laws, which are formulated with the minimum of assumptions and concepts. In the description of motion of bodies or interaction between bodies the two essential concepts are momentum and energy. These concepts are tied in with our intuitions of force and work, and the physics of motion and interaction of bodies can be described quite accurately as the study of exchange of momentum and energy. Descartes introduced the concept of momentum (product of velocity and inertial mass). He believed that this quantity was conserved but this belief was based on metaphysical-theological speculations. Huygens in a paper in 1669 first proved that momentum was conserved in an elastic collision. He also proved that the product of inertial mass and the square of the velocity of a body was conserved. The first scientist to give a detailed exposition of conservation of energy was Leibniz. He introduced this in a paper published in 1695 in which he criticised Descartes' theory of elastic collisions (which was described earlier in Chapter 3, Galileo). Leibniz called kinetic energy *vis viva* (live force), and he also introduced the concept of potential energy (at least in a gravitational field).

In the 1920s the eminent German mathematician Amalie Emmy Noether (1882–1935) realised that Newton's laws of classical mechanics had no preferred location, direction or time; that is, the laws do not depend on the origin of the temporal or spatial coordinates or on the orientation of the coordinates. To put it another way, the laws would not change (are invariant) if a constant quantity was added to the space or time coordinates. This suggested to Noether that there were three conserved quantities or laws of conservation. This discovery ushered in a new view of the deep significance of laws of symmetry or invariance. Noether was born in Erlangen and went to the University of Erlangen where she obtained her doctorate in 1907. Her thesis was on algebraic invariance. From 1913 she occasionally lectured at the University of Erlangen in place of her father who was also a distinguished mathematician. In 1915 she went to the University of Göttingen and was persuaded by the celebrated German mathematician David Hilbert (and Felix Klein) to stay there. However, she was denied a lectureship at the all-male University of Göttingen on the grounds that men should not be expected to learn mathematics 'at the feet of a woman'. Her friend and mentor, Hilbert, is supposed to have commented contemptuously, 'I do not see that the sex of a candidate is a consideration . . . after all the Senate is not a bathhouse'. Einstein regarded Noether 'the most significant creative mathematical genius'. She won formal admission as an academic lecturer at the University in 1919. In Göttingen Noether continued her mathematical research and from 1930 to 1933 she also edited the German mathematics journal *Mathematische Annalen.* When the Nazis came to power in Germany in 1933, Noether along with many other Jewish professors at Göttingen was dismissed. She emigrated to the United States, becoming a visiting professor at Bryn Mawr College, and also taught and did research at the Institute of Advanced Study, Princeton.

Conservation laws mean that certain properties of a body or system of bodies do not change – they are invariant. The concept of invariance is closely related to that of 'symmetry', which has played a fundamental role in the development of physics. The most obvious example of symmetry or invariance is a sphere because it rotates into itself. Similarly if under a symmetric operation a physical system undergoes no observable effect, the system is said to be invariant to change. Noether noticed that Newton's laws did not change under spatial displacement. This implies that for an isolated body or an isolated system of bodies in linear motion the total linear momentum of the system (that is, the sum of the product of inertial mass and velocity of all constituent bodies) remains constant in magnitude and direction. Isolated means, in this context, that there is no external force acting on the system. When a rocket is on the launch pad the total momentum of the rocket and the fuel in it is zero. In flight the downward momentum of the exhaust gases is exactly equal to the upward momentum of the rising rocket, and the sum of the momentum of the rocket and fuel system remains unchanged at zero.

The direction of a rotating body (or bodies) changes constantly and there is a centripetal force acting on the body. This body (or bodies) is not an 'isolated system' and the linear momentum of the body is not conserved. The dynamical parameter that is conserved in a rotating body is the angular momentum or the spin. This follows from the symmetry in direction or isotropy of Newton's laws. The angular momentum of a body is defined as the product of the mass and the angular velocity of the body. The law of conservation of angular momentum implies that the angular momentum of a body (or a collection of bodies) moving along a curved path does not change unless a force (torque) acts on the body (or the collection of bodies). Kepler's Second Law is an expression of conservation of angular momentum of the planets in their orbits; this was fully appreciated by Newton. A helicopter requires two propellers for stability. If there was just one propeller, on top, then the body of the helicopter would rotate in the opposite direction to conserve angular momentum.

The conservation of Earth's angular momentum keeps its rotational axis pointed towards the pole star and the length of the day stays fixed. The length of the day would only change if the Earth were to expand or contract. Over a long period the direction of Earth's axis does change but this is due to the gravitational perturbation by Sun and the Moon (more about this in Chapter 7, Hubble & Eddington). All rotating bodies, rigid or clusters of bodies, have a fixed axis of rotation or spin. The plane perpendicular to this axis and passing through the centre of mass of the body or the cluster also stays fixed: that is, its orientation does not change, Laplace called this the *invariable plane*. In a cluster of bodies if the moving bodies collide they can deflect into the invariable plane, and if they have the right angular momentum when they reach that plane they will 'settle' in it. The astronomical consequence of this is the formation of spiral galaxies and planetary systems (see Hubble & Eddington). In the solar system the spectacular planetary system of Jupiter and the rings of Saturn are a consequence of gravity and the conservation of angular momentum.

The invariance of Newton's laws to translation in time (that is, the laws do not depend on the value of the origin on the time axis) implies conservation of energy. The concept of energy was recognised by Galileo in the seventeenth century but the term 'energy' as a measure of the ability to do work (force multiplied by the distance through which the force is applied) was developed rather late. As Leibniz had recognised, energy comes in two forms: kinetic energy, which depends on the motion of a body (or system of bodies), and potential energy, which depends on the position (relative to a massive gravitating body) of a body (or system of bodies). Consider a ball thrown vertically upwards from the surface of the Earth. The ball starts with a certain speed and it has a certain kinetic energy. As the ball rises the velocity decreases and the kinetic energy of the ball decreases. This energy is lost doing work against the gravitational pull of the Earth. At the top of its trajectory the speed of the ball is zero and its kinetic energy is zero, but it has acquired potential energy and this is equal to the kinetic energy it has lost (the friction of the air is assumed to be zero). As the ball begins to fall it loses its potential energy because its height decreases, but gains speed so its kinetic energy increases. When the ball reaches the position from which it started its rise, its potential energy is zero but its kinetic energy is exactly equal to the kinetic energy with which it started. At any point in its path, the ball's kinetic energy plus its potential energy is exactly equal to the kinetic energy when the ball started its rise or its potential energy at the top of its trajectory. In this situation the mechanical energy – that is, the sum of potential and kinetic energy – is conserved. The total energy of a body (or a system of bodies) is conserved if no work is done on the body (or the system of bodies).

Conservation of energy was independently recognised by many scientists in the first half of the nineteenth century. Maxwell in *Matter and Motion*[2], published in 1876, mentions the conservation of momentum, angular momentum and energy. However, he mentions the conservation of energy as a 'generalised statement, which is found to be consistent with fact, not in one physical science only, but in all'. He does not state it as a general 'principle'. The German physicist Hermann von Helmholtz was the first to present a convincing argument for the principle of conservation of energy. Conservation of energy, kinetic, potential and elastic, in a closed system, assuming no friction, has proved to be a useful tool. Moreover, friction demonstrates itself by generating heat. During the 1840s Herman von Helmhotlz and James Prescott Joule (of England) identified heat as a form of energy. Joule also proved experimentally the relationship between mechanical and heat energy and that between heat and electrical energy. The conservation of energy is now called the First Law of Thermodynamics: it is a very general statement and applies whether the energy is mechanical, thermal, nuclear or of any other form, or a combination of these energies.

In Newton's laws of motion there is a fourth conserved quantity, namely mass. Mass enters these laws as a constant of proportionality between force and acceleration, but it embodies the fundamental concept of inertia. In Newton's laws mass is independent of position, orientation, speed or time of observations. Einstein was to show in the early twentieth century that this presumption was false and that mass, momentum and energy are interdependent quantities, related by the speed of light.

ACTION

The motion of a body can be viewed in two ways: a close-up view and the panoramic view. The close-up view involves a moment-by-moment charting of the behaviour of a body. The panoramic view reveals the overall picture of the actual motion of a body between two events and also all possible alternative routes connecting the two events. How does a body 'know' which path to follow? Does a planet look at the Sun, measure the separation, calculate the inverse square of the distance and decide to move in accordance with Newton's law? This problem was first addressed in the eighteenth century by the French biologist, astronomer and mathematician Pierre-Louis Moreau de Maupertuis (1698–1759). He became a member of Academy of Science in Paris in 1731 and helped popularise Newtonian mechanics on Continental Europe. In 1736 he led an expedition to Lapland to measure the length of a degree along a northern meridian. These measurements verified the Newtonian view that the Earth was flattened at the poles, i.e. the Earth was an oblate spheroid (more about this in Chapter 7, Hubble & Eddington). In 1744 he enunciated the principle of least action, later published in his *Essai de cosmologie* (Essay on cosmology, 1750). German mathematicians accused him of plagiarising the principle from Leibniz. In the ensuing controversy Euler supported Maupertuis but Voltaire satirised the 'earth flattener' mercilessly (and this after Maupertuis had helped Voltaire write the chapter on Newton in his *Lettres Philosophiques!*).

Maupertuis was attempting to identify the overall characteristic or property that not only favoured the correct path but also actually imposed this path on a light ray. He analysed a number of paths which a light beam could take and he found that the preferred path was one along which a quantity called the action of the light corpuscle (he assumed Newton's corpuscular theory of light) was a minimum. He defined instantaneous action as the product of momentum of the corpuscle and the change in its position. The total action along a path is the sum of all instantaneous actions. Maupertuis then proposed the following principle: the path followed by the light corpuscle is one for which the total action is a minimum. Although Maupertuis discovered the principle of least action while considering the motion of Newtonian corpuscles of light, its real importance was in its application to motion of bodies. This was first recognised by Leonhard Euler and Joseph-Louis Lagrange in the eighteenth century, who extended the action principle to mechanics. With the decline of Newton's corpuscular theory of light, the Maupertuis principle of least action as applied to propagation of light fell into disuse. The propagation of light in the wave theory of light is described more accurately by a similar 'principle of least time' discovered by Fermat (of the famous Fermat's Last Theorem). According to Fermat's principle of least time the path taken by a light beam is the one for which the time of propagation is a minimum.

Maupertuis defined his action only in terms of momentum, but the motion of a particle cannot be described in terms of momentum alone. Energy must also be taken into account. The Irish astronomer and mathematician Sir William Rowan Hamilton (1805–1865) considered this idea. In 1827 he published his paper *Theory of System of Rays* in which, starting from Fermat's principle of least time, he showed that the time or more appropriately the action can be considered as a function of the end points between which a beam is propagated. He also showed that action varied, when the coordinates of the end points were varied, according to a characteristic function (his 'law of varying action'). He showed that the entire theory of system of rays is reducible to the study of this characteristic function. In 1835, in his *On General Method of Dynamics*, he applied his idea of characteristic function to the motion of systems of bodies. He expressed the equation of motion in a form that revealed the duality between the components of momentum of a dynamical system and the coordinates determining its position. Hamilton's unification of dynamics and optics has had a lasting influence on mathematical physics, although the full significance of his work was not appreciated for almost a hundred years, until the rise of quantum mechanics.

The Maupertuis action involves only the spatial coordinates and does not account for changes that can occur in action with time, so it does not describe the general kind of motion a body can experience. Hamilton corrected this deficiency by redefining the Maupertuis action to include energy in action and the change in action with time.

The Maupertuis action is only applicable to the motion of a body in which the total energy is conserved. The Hamiltonian action is applicable in situations where the energy of the body does not have to remain constant and is therefore more general. The description of the motion of a particle in the context of the Hamiltonian action principle eliminates the need to introduce a force for the motion of a body, and the (philosophically) unacceptable 'action (of force) at a distance' can be discarded. To put it another way, the idea of causality – that a particle or a planet feels the pull of a force and moves accordingly – which is central to Newtonian mechanics is not essential.

AFTER THE *PRINCIPIA*

Publication of the *Principia* established Newton's international reputation. Initially his fame spread among the fellows of the Royal Society and a small circle of scientists. Outside this circle the full depth of his achievements was recognised rather slowly. The *Principia* was reviewed extensively; the first review was published anonymously by Halley in the Royal Society's *Transactions*, on the eve of its publication. More reviews followed in learned journals across Continental Europe. Today Newton is associated with the scientific roots of the Enlightenment: his contemporaries Robert Boyle, Robert Hooke and Christiaan Huygens tower high but Newton rules supreme. Newton's stature is summed up by Pope's often-quoted verse:

> Nature and nature's laws lay hid in night / God said 'Let Newton be!'
> and all was light.

The foundation of this recognition was his universal law of gravitation and the laws of motion. Before him Ptolemy, Copernicus, Brahe and Kepler had attempted, in their different ways, to understand the motion of planets. Galileo had introduced the methodology of experiments. But it was Newton who had welded together the motion of planets and Galileo's idea of force to write down the equations that describe the paths of planets and of falling bodies. He extended the rule of physical laws to the entire solar system and by implication to the universe. It was Newton who finally broke the hold of theology on the cosmos and led to the belief in a rational explanation for the workings of the universe.

After the completion of the *Principia*, Newton's life began to branch away from the academic isolation in which he had lived for 25 years. The changes were brought about by his altered perception of himself and the by changes that were taking place outside the University of Cambridge. King Charles II died in February 1685 and his Catholic brother James II ascended the throne of England. James decided to change the rules of the universities to allow Catholics onto degree courses and to hold high offices. The strongly Protestant Establishment of Cambridge decided to fight back and Newton was appointed to the committee of eight representatives to plead the case for the University. Newton threw himself into preparing this case. James brushed aside all the arguments that were put to him and pushed through his reforms. However, by 1688 James was in exile in France and in December that year William of Orange was welcomed as King of England. The dispute with James had highlighted Newton's ability to marshal arguments and forcefully put a case for strongly held beliefs. The university authorities noticed this. Newton had also acquired a taste for official responsibility and he was becoming aware of operating within a wider political and social world. Newton represented the University of Cambridge in the Convention Parliament convened to transfer power from the Stuarts to the House of Orange. Newton MP made absolutely no contribution to the proceedings of the new parliament but he came into contact with and made friends with influential people. Amongst these were Charles Montagu, later to become the Chancellor of Exchequer, and the English philosopher John Locke.

Newton had been concerned about the future direction of his career after his year as an MP. In 1695 Charles Montagu, by then appointed Chancellor of Exchequer in the Whig government of 1694, offered Newton the post of Warden of the Royal Mint and Newton left Cambridge for London. Remarkable as it may seem, the Royal Mint was broke in the late seventeenth century. The principal cause of loss of revenue was the practice of 'clipping' – cutting the edge off coins, the clipped silver being sold separately at a higher price. The Mint continued to give the full exchange value of the clipped coins. Newton introduced coins with a milled edge as part of wide-ranging reforms at the Mint, coins without a milled edge having no redeemable value. Newton proved to be a perfect administrator, bringing the skill and intellectual rigour he had displayed in preparing the *Principia* to the problems of recoinage. Newton seems to have toyed with the idea of decimalisation of the British coinage, but this reform was deferred for almost 275 years. In 1699 Newton was appointed the Master of the Mint upon the death of the incumbent Master, Thomas Neale. Perhaps for Newton a more important development was to be elected President of the Royal Society on 30 November 1703. This was a turning point in the fortunes of the Society. Here, again, Newton introduced sweeping reforms. To control the funding crisis which had plagued the Society for some time, he introduced admission fees for Fellows, and sold the Society's stock in the East India Company and the East Africa Company. By 1710 the financial position was transformed allowing Newton to move the Society to new and larger premises (and in the process lose Robert Hooke's portrait!). The Royal Society was saved not only by Newton's impressive administrative ability but also by the iron discipline he introduced in matters scientific. This re-established the Society's scientific respectability. As President of the Royal Society Newton assumed the role of the patriarch of English science and ruled the Society magisterially (or tyrannically, according to John Flamsteed, the first Astronomer Royal).

In 1704 Newton presented to the Society his second great scientific work, his book *Opticks*. Unlike the *Principia* this was written in English and contained very little mathematics. The *Opticks* was originally conceived of as four books. The last book was to be a description of a grand unification of the optical phenomena in the first three books with the mechanical theories described in the *Principia*, a theory to bring together all the known forces of nature. This appears to have been Newton's aim even when he was preparing the *Principia*. In the

'Conclusio' to the *Principia*, which was never published, Newton states that like gravity, electricity and magnetism there must be other forces in Nature as yet unobserved, and just as large bodies act upon each other there must be forces that act upon 'insensible particles'. Newton was unable to achieve this grand unification, which remains, even today, the 'Holy Grail' of physics. In 1705 he became the first scientist to be knighted. Newton died on 20 March 1727 and on 4 April was buried in Westminster Abbey (alongside the monarchs of England), in the part of the abbey that has since become known as Scientists' Corner – a place of rest for illustrious British scientists.

Behind the towering genius of Newton, there was a man of flawed character. The death of his father before his birth and the separation from his mother at an early age had left deep scars on his personality. His psychotic tendencies have been ascribed to these traumatic events. Women do not seem to have played a significant part in his life and are not mentioned in any of his writings. However, he had a well-developed sense of self-preservation and personal advancement, and he was very careful to cultivate friendships that helped him advance his career. His deep sense of insecurity rendered him obsessively anxious when his work was published and irrationally violent when he had to defend it. A sad example of this is the dispute with Leibniz over the priority for the invention of calculus. Sometime between 1666 and 1670 Newton invented the new mathematics of fluxions (now known as calculus) to deal with instantaneous motion. He applied this to his theory of gravitation to calculate the paths of planets under the influence of an inverse square law of force. In the *Principia* Newton hinted at this method, but did not really publish it until 1704 when two papers were appended to the *Opticks*. By then the priority controversy was in full blaze.

The German philosopher and mathematician Gottfried Wilhelm von Leibniz (1646–1716) was born in Leipzig and grew up in an atmosphere of strict piety, and he maintained this religious outlook throughout his life. In 1672, during a visit to Paris, Huygens introduced him to his work on the theory of curves. Encouraged by Huygens, Leibniz immersed himself for next several years in the study of mathematics. Between 1673 and 1675, working in complete isolation, he produced a new canon of mathematics which included the method of infinite series and, more importantly, a version of calculus. He stressed the power of his calculus to investigate 'paths' of a class of 'mechanical objects' which Descartes had believed lay beyond the power of analysis. In 1684 Leibniz published his first paper on calculus in *Acta Eruditorum*, a learned journal produced by the University of Leipzig. This paper brought calculus to the attention of other mathematicians. When Leibniz was informed that the priority for invention of calculus might be attributed to Newton, he responded with characteristic modesty:

> As far as Mr Newton is concerned, I have a letter from him and Mr Oldenburg in which they do not dispute my quadrature with me, but grant it... Meanwhile, I acknowledge that Mr Newton already had the principle from which he could well have derived the quadrature, but one does not come upon all the results at one time: one man makes one contribution, another man another.

But the notion of parallel invention was unthinkable for Newton and he defended his claim with passion. What began as mild innuendo rapidly escalated into blunt charges of plagiarism. Even Leibniz's death could not allay Newton's wrath, and he continued to pursue his adversary beyond the grave. The battle with Leibniz dominated the last 25 years of Newton's life. National chauvinism fuelled this dispute over priority into a battle of ideologies. This eventually developed into a schism in philosophical thought and mathematical practice between England and Continental Europe and lasted for generations. The superior notation, developed by Leibniz, was adopted by European mathematicians but was deliberately ignored by their British contemporaries who consequently lost the advantage they had in mathematics. Leibniz's notation has now become part of the mathematical language and is universally used. This feud between Newton and Leibniz was largely pointless and there is certain tragedy in Newton's reluctance to acknowledge the superiority of Leibniz's analysis. Newton is rightly seen as the first man to have invented calculus but for a limited application, to obtain the paths of planets. Leibniz developed calculus as an intellectual exercise and had a far broader application in mind. Mathematics would have looked much the same after Newton, but Leibniz and the community of researchers which his analysis attracted enormously influenced the subsequent developments in mathematics and mechanics.

Newton was obsessed with alchemy, Biblical prophecy and strange metaphysics. His alchemical and religious writings far exceed his scientific publications. In 1690 he sent to the British philosopher John Locke a copy of a thesis attempting to prove that the Trinitarian passages in the Bible were a latter-day corruption. When Locke proposed publication Newton withdrew the thesis fearing that his Arian views would become known. In his later years, he devoted much time to the interpretation of Biblical prophecies and the study of Biblical chronology[3]. Towards the end of his life Newton burned a large quantity of papers pertaining to his works on mystical subjects and his mystic life has not received much attention. Underneath his scientific writings there is a belief in a harmoniously ordered cosmos, a manifestation of an underlying mystic order. It is impossible to tell how far the *Principia* and the *Opticks* were influenced by inspirations he derived from his alchemical experiments. But to his lasting credit he did not allow his alchemical speculations to detract him from the strict mathematical rigour and experimental verification he employed in his scientific research.

Newton's scientific achievements represent the peak of the Scientific Revolution (the series of rapid changes in our knowledge of nature and the mode of studying natural phenomena) that began in the late sixteenth century. The core of Newton's great revolution is his establishment of the science of mechanics on mathematical principles. His introduction of the modern concept of mass is essential for the study of matter. Newton is best known for his discovery of the principle of universal gravitation. He not only discovered the principle but also formulated the quantitative laws of gravity. His laws of motion and the laws of gravity make it possible to arrive at a logical explanation of tides, the motion of comets, precession of the Earth's axis and the motion of any body in a gravitational field (man-made satellites and space probes, for example). Although this theory synthesises celestial and terrestrial motions, Newton had realised that the theory was not complete in the truly physical sense, as he was unable to explain what gravity was. In the *Principia* he states clearly:

> I design here to give a mathematical notion of those forces, without considering their physical causes and seats . . . the reader is not to imagine that . . . I anywhere take upon me to define the kind, or the manner of any action, the cause of the physical reason thereof.

These statements encapsulate the spirit of the *Principia*: Newton attempts to explain 'how' forces influence bodies and not 'why' they are so influenced. He puts it succinctly at the conclusion of the *Principia*:

> *hypothesis non fingo* (I feign no hypothesis).

Newton's work had a dramatic effect upon an entire culture. He was not only a genius but was popularly and universally perceived as such. The nature of the Newtonian revolution was not at first apparent, but the scientific importance of the *Principia* (as of his other scientific contributions) was immediately recognised. Newton wrote the *Principia* for mathematicians and is beyond the comprehension of most. But this was soon put right. In France the task of interpreting and popularising Newtonian philosophy was undertaken by the writer Voltaire (pen name of Francois-Marie Arouet 1694–1778). Voltaire was aided by his mistress Émilie de Breteuil, Marquise du Chastelet (1706–1749). The Marquise was one of the very few women mathematicians of the eighteenth century and had translated the *Principia* into French (published posthumously in 1759). In 1734 in his *Philosophical Letters* Voltaire introduced Newton as the 'destroyer of the system of Descartes'. His authoritative and delightfully lucid exposition marks the real triumph of the Newtonian revolution. The impact of Newton's work was such that it was considered essential to understand it, starting a trend of 'public understanding' of science. A number of popular accounts of Newton's work were produced – some profound, others banal. But as Voltaire has commented:

> Newton had very few readers because it requires great knowledge and sense to understand him. Everybody however talks about him.

Not unlike Stephen Hawking, the current Lucasian Professor of Mathematics, and his book *A Brief History of Time*.

5 Einstein

For 200 years, from the beginning of the eighteenth century to the beginning of the twentieth century, Newtonian mechanics reigned supreme. By the late nineteenth century, the few simple laws of Newtonian physics could explain, with uncanny accuracy, most of the disparate phenomena of the natural world. Everything in the heavens and on Earth appeared to obey the laws, and the mastery of these laws was bringing mankind the mastery of the environment. These laws dominated the way both scientists and laypersons thought. Newtonian mechanics has a deterministic framework for the cosmos and this was deeply satisfying to the Judeo-Christian culture of Western Europe. There were those who questioned Newton's assumptions of absolute space and time, independent of man, but anyone doubting the validity of the laws of motion or gravitation was not taken seriously by the scientific community.

Newton had applied his theory of gravitation to two-body systems, such as the Sun and a planet. In the eighteenth century various attempts were made to extend it to three gravitating bodies. In 1682 Halley had claimed that the comet then observed in the sky had also appeared in 1531 and 1607; given, then, that the period of the comet was about 75 years, he predicted that it would reappear in 1758. Months before its appearance, the French mathematician Alexis Clairaut used tedious and brute-force mathematics to calculate the gravitational perturbations of Jupiter and Saturn on the otherwise elliptical orbit of Comet Halley. Clairaut predicted that the comet would reach perihelion in April 1759, give or take a month. Its appearance in March was an early confirmation of the scope and power of Newton's theory.

Furthermore, elegant and general solutions were found for three-body problems involving two planets and the Sun, or the Sun, a planet and its moons. Because the orbits of these bodies were in the same plane, and they described nearly circular orbits in this plane, simplifying assumptions were possible. In the nineteenth century, Lagrange and Hamilton reformulated classical mechanics within a very general mathematical framework. Their equations form the starting point of most modern applications of the subject. They applied the calculus of variation, as developed by Leibniz, to attack astronomical problems. It is a delightful irony that the calculus of Leibniz should have been responsible for deepening our understanding of the mechanics developed by his adversary, Newton. It is also amusing that Leibniz himself had grave doubts about the theory of gravitation. He was particularly concerned about its 'action at a distance', which he thought introduced occult forces into physics (see Chapter 4, Newton).

The great mathematician of the eighteenth century, Giuseppe Luigi Lagrangia (1736–1813), more often known by the 'French version' of his name, Joseph-Louis Lagrange, undertook a detailed analysis of the Moon's motion. It was known since Galileo's time that while the Moon always has the same face to Earth it also appears to oscillate so that features near the edges are alternately visible and invisible. This is known as the libration of the Moon. Lagrange showed that this was a result of the slightly nonspherical shape of both the Earth and Moon. The gravitational attraction of these bodies could not be assumed (as Newton had done) to be proportional to the distance between the centres of these two bodies. This won him the Paris Academy of Science prize in 1764. He also won this prize in 1766, 1772, 1774 and 1778 for finding solutions of various problems in mechanics. In his classical work, *Mécanique analytique*, Lagrange brought together a masterly synthesis of almost one hundred years of research in mechanics since Newton. These days Lagrange is better known for the set of 'neutral' points he discovered in a system of two gravitating bodies. A neutral point is a point where the gravity of the two bodies cancels exactly. In a system of two orbiting bodies where one body is much more massive than the other (the Sun and the Earth, say) the light body will have an elliptical orbit round the heavy body. If the effects of motion are ignored then there will be a single neutral point. However, in a real two-body system, apart from the gravitational force of the two bodies there is also a force associated with the motion of the body in orbit, namely the accelerating centripetal force. Lagrange discovered that the gravity of the two bodies and the centripetal force are in equilibrium at five locations called Lagrange or libration points (after the Greek word *Libra* – balance) designated as L_{1-5}. In the Earth–Sun system a satellite placed at a Lagrange point will experience minimum perturbing forces from Sun and Earth. Several recent (and proposed) space probes have taken advantage of this unique property of Lagrange points for long-duration space missions. The joint European Space Agency/NASA solar mission SOHO (Solar and Heliospheric Observatory) launched in December 1995 is hovering near the L_1 point, about 1.6 million kilometres from Earth in the direction of the Sun. This is a perfect vantage point for observing activity on the Sun and relaying the information to Earth.

A contemporary of Lagrange, Pierre-Simon Laplace (1749–1827), applied Newton's theory of gravity to the entire solar system to analyse the problem of the orbits of Jupiter and Saturn. This problem had been noted as long ago as 1650, and Halley had also commented on it. The troublesome problem was that Jupiter's orbit appeared to be continuously shrinking while that of Saturn appeared to be expanding. Laplace established, in 1773, that the mean motion of planets was invariant or unchanging. This was the first and the most important step in establishing the stability of the solar system. Laplace, working on suggestions by Lagrange, showed that the changes in the orbits of Jupiter and Saturn were not secular but periodic with a period of 926 years. Laplace followed this to show that the mutual gravitational effects on planetary orbits were self-correcting. In a five-volume *Traite de mécanique celeste* (Celestial Mechanics), published between 1798 and 1827, Laplace presents a complete mechanical interpretation of the solar system, the motion of planets and their satellites, and their perturbations. With the publication of this work the Newtonian problems of planetary motion appeared to have been solved.

In 1796 Laplace published *Exposition du systeme du monde* (The system of the world), a semi-popular treatment of his work on celestial mechanics. This work includes his 'nebular hypothesis' – Laplace pointed out that the motion of all members of the solar system is almost circular, almost in the same plane and also in the same direction. This suggested to him that the solar system might have condensed out of a vast rotating mass of gas, a huge gaseous nebula. This hypothesis has had a strong influence on the subsequent development of theories of the origin of solar systems (see Chapter 7, Hubble & Eddington).

Both Lagrange and Laplace were spared the political and social upheaval in France in the last quarter of the eighteenth century. Both Louise XVI and the Revolution respected them and their work. Their colleague, the great chemist Antoine-Laurent Lavoisier (1743–1794), was not so fortunate; in 1794, he was guillotined by the Revolution. Lagrange commented sadly that 'It required only a moment to sever that head, and perhaps a century will not be sufficient to produce another like it'.

The most impressive observational astronomer of the eighteenth century was William Herschel (1738–1822). Born in Hanover, Germany, he came to England in 1757 as a musician. Fortunately he turned to astronomy and acquired great skill in making instruments. He conducted four complete surveys of the sky with telescopes of increasingly greater power. The second survey in 1781 revealed Uranus, the first planet discovered in historic times. Further improvements in his instruments resulted in the discovery of moons of Saturn and Uranus. His survey of the stellar systems led him to conclude that stars were distributed in a lens shape, the edge being formed by the Milky Way. The diameter of the lens was about five times its thickness and the Sun was close to the centre of the lens. In 1805 Herschel discovered that some stars circulated around each other. In their binary motion these pairs followed the law of Newtonian gravitation. Thus the laws that Newton had formulated to explain the solar system were demonstrated to apply also to the distant stars.

Around the middle of the nineteenth century Newtonian theory was applied in a much more dramatic manner to make a discovery which captivated the imagination of the public. Before the discovery of Uranus by Herschel in 1781, the consensus of opinion among scientists was that the only planets in the solar system were the six that had been known since antiquity. The discovery of the seventh planet led astronomers to suspect the existence of still more planets. Careful observations of the orbit of Uranus led to the discovery that the path did not agree with the orbit calculated from Newtonian laws.

In 1843 the British mathematician John Couch Adams (1819–1892) began a careful analysis of the path of Uranus and the path predicted under the gravitational influence of the Sun and the known planets. Adams concluded that a more distant planet was perturbing the path of Uranus. He communicated his results to Airy, the Astronomer Royal, but Airy was out of town and did not see the results for some time. John Herschel, son of William Herschel, convinced Airy of the possibility of discovering a new planet and in 1846 Airy sent Adams' calculations to James Challis at the Cambridge Observatory. Challis began a careful search of the area of the sky surrounding the position predicted by Adams. The search was slow and tedious; Challis did not have catalogues of the faint stars in the area of the sky where the new planet was predicted. Challis drew charts of the stars he observed and then compared these with the same region a few days later to see if any star had 'moved'.

Around this time Urbain-Jean-Joseph le Verrier (1811–1877) in France, unaware of Adams' work in Cambridge, started a similar study of the path of Uranus. Le Verrier had difficulty convincing astronomers in France that it was worth spending telescope time to search for a new planet. He communicated his results to Johann Gottfried Galle at the

Berlin Observatory. Galle and his assistant made a careful search of the charts made to aid the identification of asteroids. They immediately identified Neptune and verified the identification the next night from its motion relative to the background stars. The discovery of Neptune at the predicted position was a widely understood confirmation of Newtonian theory. In 1915 the American astronomer Percival Lowell published predictions of yet another planet to account for perturbations of Uranus not accounted for by Neptune. Pluto was discovered in 1930 by the then new technique of photography.

SPEED OF LIGHT

Despite its successes it is sobering to realise that the seed of a challenge to Newtonian mechanics had been sown 10 years *before* the publication of the *Principia*. In 1676 Roemer announced to the *Académie des Sciences* in Paris that he had measured the speed of light – the concepts of absolute and independent space and time were dead. Olaus or Olaf Christensen Roemer (1644–1710) went to Paris in 1672 to work at the new observatory set up under the directorship of Giovanni Domenico Cassini (1625–1712), who was called to Paris by Louis XIV in 1669, and became the most influential figure in French astronomy. Contrary to his time, Cassini was anti-Copernican. He was succeeded at the observatory by three generations of his descendants and the Cassini regime at the observatory lasted for over 150 years. Their anti-Copernican views gradually weakened as the dynasty came to an end but it was injurious to French science. Cassini measured the distance to Mars by making simultaneous observations, against a background of fixed stars, from Paris and Cayenne in French Guiana. The distance to Mars could then be obtained by the simple method of triangulation used in surveying. From this measurement Cassini obtained the Earth–Sun distance (the Astronomical Unit, AU) by Kepler's Third Law. The result was 140 million kilometres, only 7% lower than the true value of 150 million kilometres.

When Roemer came to the Paris Observatory he decided to take a look at the extensive observations of Jupiter made by Cassini. He noticed that the time of occurrences of eclipses of the satellites of Jupiter varied with season. Roemer deduced correctly that this was because the speed of light was finite and not, as had been previously assumed, infinite. If the Earth was stationary in its orbit then an observer on Earth would notice that every eclipse started after the same interval. But if the Earth is receding from Jupiter then successive eclipses will be separated by longer intervals because the light from these eclipses will have to travel an additional distance to reach an observer on Earth. Roemer found that the delay when Earth was closest to and furthest from Jupiter was 1000 seconds. This delay represents the time taken by light to travel across the diameter of the Earth's orbit, that is, 2 AU. This suggested that the speed of light is 0.225 million kilometres per second (the modern value is 0.2998 . . . million kilometres per second). This measurement was of monumental importance, but the conservative Cassini rejected it and it was ignored. It was not until 1728, when the British astronomer James Bradley confirmed Roemer's measurement, that the finite speed of light was widely accepted. In 1681 Roemer returned to Denmark where he was appointed Royal Astronomer and a professor of astronomy at the University of Copenhagen.

Laboratory studies of the nature of light continued in the eighteenth and nineteenth centuries. Building on the earlier work of Huygens, Thomas Young (1773–1829) established the wave nature of light. It was also realised that, because of its wave nature, the wavelength and hence the frequency of light would be modified if the source of light was in motion relative to an observer. If the source moved towards an observer, the waves would be squashed up at the observer, and the wavelength would appear shorter (Figure 5.1), so light would appear bluer. If the source moved away from an observer, the waves would be stretched out at the observer (Figure 5.1), the wavelength would appear longer and light would appear redder. Thus the relative motion of a source and an observer results in a blue or a red shift of the light. In 1842 the Austrian scientist Christian Doppler (1803–1853) predicted this change in the wavelength or the frequency of light when the source and

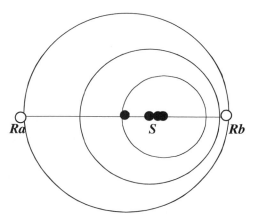

FIGURE 5.1 Doppler effect. The circles can be considered to be crests of light or sound waves emitted from the moving source S. At receiver Rb, towards which the source is moving, the crests are bunched up or the wavelength is shortened and the wavelength of light is shifted to the blue. At receiver Ra, from which the source is receding, the crests are pulled out or the wavelength is longer and the wavelength of light is redshifted. The effect is most readily observed when a train passes by a passenger on a platform; as the train approaches the passenger the pitch of the train's whistle increases (the frequency gets higher), and as it recedes the pitch gradually decreases (the frequency decreases).

observer are in relative motion, and this is now known as the Doppler effect. The first experimental verification was made by Buys Ballot in 1843. Christophorus H. D. Buys Ballot (1817–1890) was a Dutch meteorologist and is better known for his observation in 1857 that wind blows at right angles to the atmospheric pressure gradient (he was not the first to notice this, but his name is associated with it). He also made important contributions to the kinetic theory of gases. He demonstrated the Doppler effect by mounting trumpeters on a railway carriage and measuring the pitch of the notes as the carriage was moved past an 'observer'.

WHAT IS A STRAIGHT LINE?

By about the middle of the nineteenth century mathematicians were also beginning to question the basis of Euclidean geometry, in particular his parallel axiom (see Chapter 1, Aristotle). As early as 1817 Carl

Friedrich Gauss (1777–1855) suspected that Euclidean geometry was mathematically incomplete. He suspected that the parallel axiom was not really an axiom (that is, assumed independently) but a theorem and that it should be possible to prove it from other Euclidean axioms. But all the proofs of the parallel axiom contained errors, and Gauss realised that there might be a geometry in which the parallel axiom was not valid, yet that geometry would be internally consistent and free of contradictions. But Gauss did not publish his work! Gauss was a deeply conservative man, in his politics, his personal life and in his mathematics. He did not once travel outside Germany and spent almost his entire life in one city. It was not for him to break out of the three-dimensional Euclidean geometry and his theory had to be reconstructed after his death. Other geometries were constructed in the nineteenth century, the outstanding ones being those of the Russian Nicholas Lobachevsky (1792–1856), the Hungarian Janos Bolyai (1802–1860) and the German George Bernhard Riemann (1826–1866). Riemann was one of the very few students of Gauss. His general geometry of space with three or more dimensions was the subject of his inaugural lecture at Göttingen in 1854 and is said to have agitated Gauss. At 28, he published a paper entitled *On the Hypotheses which lie at the Basis of Geometry*. The ideas therein demonstrated that the Euclidean geometry was only one of many possible geometries. Riemann's essay forms the basis of our present conception of space-time (described later in this chapter). In many ways Riemann was perhaps the least likely person to overthrow a subject which appeared to be solidly built on Greek foundation. He was born in 1826 in Hanover, Germany, the second of six children of a German pastor. His father, an ex-soldier, struggled to bring up his large family on the poor salary of a country cleric. Malnutrition was responsible for the early death of most of the Riemann children and of their mother. Young George was a frail and a shy child who suffered repeated bouts of illness (he suffered from tuberculosis) and bouts of nervous breakdown. Nothing in his personality suggested the breath-taking boldness of vision that he brought to his scientific work. But at an early age Riemann demonstrated his fantastic ability with numbers and calculations. To please his father, Riemann studied theology, his aim being to get a paying position to help support his family. Although he concentrated on his Bible studies, his thoughts kept drifting off to mathematics. At school his amazing ability to grasp complex mathematical ideas was recognised. His far-sighted father, ignoring the poor state of his family's finance, scraped together enough funds to send Riemann to the famous University of Göttingen. At 19 Riemann started his studies under the towering figure of Gauss. But life was never kind to Riemann; just when he thought his fortunes had taken a turn for the better, a full-scale revolution swept through Germany. In the insurrection of 1848 all of Germany was in turmoil and Riemann was forced to interrupt his studies as he was inducted into the armed forces.

The break with Euclid's geometry came when Gauss asked his student Riemann to prepare an oral presentation on the 'foundation of geometry'. Riemann was horrified; he had a pathological fear of public speaking. But over the next several months he started developing a theory of higher dimensions. During this time he was also helping another professor, Wilhelm Weber, with experiments in electricity, a fascinating new field at the time. Riemann was excited by the new discoveries being made in electricity and magnetism and he was convinced that it would be possible to unify electricity and magnetism mathematically. However, burdened with the preparation of the public lecture, the electricity experiments and concerns about his family, his health collapsed and he suffered a nervous breakdown in 1854. After spending several months recovering, Riemann finally delivered his talk and it was received enthusiastically. The 10th of June 1854 was a decisive day in the history of mathematics; the elegant structure of Euclidean geometry had been swept aside. News of the lecture soon spread throughout Europe.

The difference between Riemann's geometry and that of Euclid can be illustrated by Riemann's axiom that all lines are finite but endless. There is no contradiction here; if you consider a sphere, any line drawn on the sphere will return to the point of origin, so the line is finite in length but there are no end points. Riemann's mathematics was wonderfully suited for the future development of the theory of relativity. This is hardly surprising, as there is evidence that Riemann was himself attempting a unified description of light, electromagnetism, heat and gravitation. He realised that if a body attempted to move across a curved surface it would appear to feel a 'force' which would prevent it from moving in a straight line. Riemann had thus arrived at a momentous conclusion: *force can be a consequence of geometry*. This was a profound break with Newtonian mechanics.

In the Euclidean geometry the definition of a straight line is self-evident; it is the shortest distance between two points. But in non-Euclidean geometries a straight line cannot be defined in such a simple way. Consider a sphere again: the straight line between two points on a sphere is a segment of a great circle through the two points. (The shortest distance would be a hole drilled between the two points, but this is not allowed.) A great circle is the circle formed by intersecting a sphere with a plane that passes through the centre of the sphere. On Earth, the meridian circles are great circles but latitude circles are not, except, of course, for the equator. Planes fly along great circles. The technical term for these great circles on a sphere is *geodesic*. A geodesic can be defined as the least curved line that can be constructed on a (curved) surface, and is the shortest curve connecting any two points. On a cylinder, the shortest distance between two points (not on a line parallel to the axis of the cylinder) will, in general, be a segment of an ellipse. It is now easy to see why Euclid's parallel axiom fails in non-Euclidean geometries. Consider a sphere (once again!) and draw a pair of parallel lines perpendicular to the equator: according to Euclid's axiom these two lines should not meet when they are extended. But on a sphere a straight line is a great circle, and the two lines, parallel at the equator, when extended along the sphere will meet at the poles, in violation of Euclid's axiom. The definition of a straight line thus depends on the geometry of the surface.

The non-Euclidean geometry of a surface can also be distinguished by considering the properties of standard geometrical figures on these surfaces. For example, on a Euclidean surface the sum of angles of a triangle

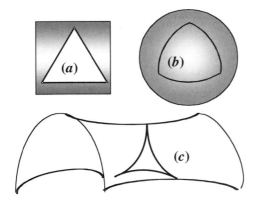

FIGURE 5.2 Geometries with zero (a), positive (b) and negative (c) curvatures. On a surface with zero curvature the sum of angles of a triangle is 180 degrees, for a surface with positive curvature the sum is greater than 180 degrees and for a surface with negative curvature the sum is less than 180 degrees.

is 180 degrees; the sum of angles of triangles of smaller or larger dimensions will also be 180 degrees. Such a surface is said to have zero curvature (Figure 5.2(a)). But the sum of angles of a triangle on a spherical surface will be greater than 180 degrees (Figure 5.2(b)). As the size of a triangle on this surface is increased the sum of the three angles increases and can equal 270 degrees. A surface like this is said to have a positive curvature. A surface with a negative curvature is saddle-shaped (Figure 5.2(c)). The sum of the angles of a triangle on this surface is less than 180 degrees and the sum varies with the size of the triangle. But a very small section of a surface with a positive or a negative curvature can look like an Euclidean surface.

Riemann had developed a powerful technique for describing space of any dimension with arbitrary curvature. To everybody's surprise there were none of the expected contradictions and the Riemannian spaces were well defined and self-consistent. Riemann continued his work in mathematics and physics and in 1858, he even announced that he had finally succeeded in unifying light and electricity. But he was unable to develop the field equations of the forces he was attempting to unify. Unfortunately Riemann's success in mathematics and physics did not translate into money. After many years he was appointed to Gauss's position at Göttingen, but it was too late. In 1866 Riemann died of consumption at the relatively young age of 39.

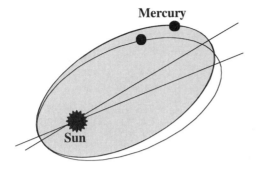

FIGURE 5.3 A schematic representation of the orbit of Mercury. The long axis of the orbit rotates in the plane of the orbit. The rate of precession is 43 seconds of arc per century.

DOUBTS ABOUT NEWTONIAN MECHANICS

The nineteenth century was the 'high noon' of Newtonian mechanics. Great mathematicians of that era were reformulating the laws of Newtonian mechanics into their most elegant form. There was also rapid progress in the study of chemical, electrical and optical properties of matter. From these studies the sciences of chemistry, electricity, magnetism and optics were to develop. Towards the end of the century doubts were beginning to be cast about Newtonian ideas of motion and gravitation. The doubts were both conceptual and observational and can be summarised under three headings: (1) the motion of the planet Mercury, (2) the concept of action at a distance and (3) symmetry between mechanical and optical (accurately electromagnetic) phenomena. It is worth considering these in detail.

Kepler's careful analysis of data obtained by Tycho Brahe had established that the orbits of planets were elliptical. Newton provided the theoretical background for these observations based on his laws of motion and gravitation. Observations of planetary orbits, made over the following several decades, suggested that there were minor discrepancies in the orbit of the planet Mercury. Mercury is the innermost planet of the solar system. Its distance from the Sun is 57.9 million kilometres and in the solar system, it has the most eccentric orbit. The period of the planet is nearly 88 Earth-days. The general geometry of the orbit is as shown in Figure 5.3. The orbit of Mercury is, however, anomalous in the sense that the point of closest approach to the Sun, the perihelion, shifts with each orbit. The rate of advance is very slow – the line joining the perihelion with the Sun turns through an angle of 575 seconds of arc (i.e. about a sixth of a degree) in 100 years. Detailed calculations of the orbit of Mercury including the perturbations of other planets accounted for 532 arcseconds of this advance. Most of the perturbation is due to Venus (the planet closest to Mercury) and Jupiter (the most massive planet in the solar system). The residual advance of 43 arcseconds per century could not be accounted for by Newtonian mechanics.

By the middle of the nineteenth century this problem of the perihelion of Mercury had become something of a challenge for astronomers. Le Verrier (later to be the co-discoverer of Neptune) reasoned that the perturbation of the orbit of Mercury was due to an undiscovered planet located between Mercury and the Sun. He provisionally named this undiscovered planet *Vulcan*. However, very careful searches failed to find this putative planet and the advance of perihelion of Mercury remained unexplained. The unexplained discrepancy was only 8% of the total advance of perihelion but the precision of observations was enough to cast doubts on the validity of Newton's law of gravitation.

By the end of the nineteenth century the *action at a distance* of Newtonian gravitation was recognised to be unintelligible. An imaginary scenario[1] perhaps best illustrates this. The Earth moves in an orbit around the Sun and the Earth–Sun distance is 150 million kilometres. Suppose the Sun is suddenly removed completely from this system. Because of gravity's instantaneous action at a distance, the Earth will be free of the Sun's gravitational attraction and would *immediately* begin to move out of its orbit, at about 30 kilometres per second. This could be verified by observations of the background stars. But the speed of light is about 0.3 million kilometres per second so although the Sun has been removed from the Earth–Sun system *the Sun will be visible for the next 8 minutes*. For eight minutes the Sun would be visible but the Earth would behave as if it did not exist. This is not logical!

Newton formulated laws of motion and gravitation to study the effect of mechanical force on massive bodies. But there are other forces in nature, which were also being investigated in eighteenth and nineteenth centuries. One such force under very active investigation was the electrostatic force. In 1785 the French scientist Charles-Augustine Coulomb (1736–1806) quantified the laws of interaction between electrically charged bodies. He used a torsion balance (described in the next chapter) to measure the electrical force between two charged pithballs. (It is interesting that Cavendish had conducted this experiment in 1771 but did not publish his results, which remained unknown till 1879). Coulomb demonstrated that the two electrostatically charged bodies attract (or repel) with a force that varies precisely as the inverse square of the distance separating the bodies. He also demonstrated that the attraction (or repulsion) is proportional to the product of the charge on the two bodies. The similarities between Coulomb's laws of electrostatics and Newton's laws of gravitation were striking. The difference was that gravitation only attracts while in electrostatics both attraction and repulsion is possible.

The other force being investigated in the nineteenth century was that of static magnetism. This displayed a number of similarities with electrostatic forces. Faraday experimentally investigated the nature of electric and magnetic forces. Michael Faraday (1791–1867) was born in a poor family and at the age of 13, after a nominal education, was apprenticed into the bookbinding trade. This proximity to books seems to have fired his interest in learning and he rapidly became impressively self-educated. He also became interested in experimental science and attended lectures by Sir Humphry Davy at the Royal Institution. The comprehensive notes he made of these lectures got him a job, in 1812, at the Royal Institution, as a laboratory assistant to Davy. There Faraday began his career as a chemist. He wrote a manual of practical chemistry that reveals his mastery of the technical aspects of his art and he also discovered a number of new organic compounds. In a few years he also became an exceptional experimental physicist and his major contribution was in the field of electricity and magnetism. He was the first person to produce an electric current from a magnetic field, invented the first electric motor and dynamo, demonstrated the relation between electricity and chemical bonding, and discovered the effect of magnetism on light. A Fellowship of the Royal Society was inevitable and he eventually became the Director of laboratories of the Royal Institution. Faraday also had a gift for explaining science in a simple language and he became a brilliant populariser of science. A profoundly modest and retiring man, Faraday refused many public honours that were offered to him.

In 1820 Hans Christian Ørsted (1777–1851) had shown that the flow of electricity in a wire produced a magnetic field around the wire. André-Marie Ampère (1775–1836) showed that the magnetic force around a current-carrying wire was circular, producing in effect a cylinder of magnetism around the wire. No such circular force had ever been observed before. Faraday understood the significance of these discoveries; if a magnetic pole could be isolated, it ought to move constantly in a circle around a current-carrying wire. This was to lead to the invention of the electric motor, but equally importantly it led Faraday to contemplate the nature of electricity. Unlike his contemporaries, he was not convinced that electricity was a material fluid that flowed through a wire like water though a pipe. Instead he believed it was a vibration or force that was somehow transmitted as a result of tensions created in the conductor. In 1831 Faraday showed that when a permanent magnet was moved in and out of a coil of wire a current was induced in the coil. Magnets, he knew, were surrounded by lines of force that could be made visible by sprinkling iron filings on a piece of paper held over them. Faraday interpreted these 'lines of force' as lines of tension in the medium surrounding the magnet and he showed that the magnitude of current in the coil was proportional to the number of lines of force cut by the conductor per unit time. This was to lead to the invention of the dynamo. These observations led Faraday to propose an underlying unity of the forces of nature. By this he meant that all the forces of nature were manifestations of a single universal force. In 1846 he made these speculations public, specifically referring to point atoms and their infinite fields of force. He suggested that the lines of electric and magnetic force associated with these atoms might serve as a medium by which light waves were propagated. But by 1850 Faraday had evolved a radically new view of space and force. Space was not the mere location of bodies, but a medium capable of supporting the tensions of forces. Energy was not localised in particles from which forces arose but rather was to be found in the space surrounding the particles. Thus was born the field theory.

Changes in the magnet or the electric charge could now be considered to create a wave in the surrounding field that would carry the change to a distant point, rather like the waves that spread along the surface of a pond when a stone is dropped in the pond. The waves would persist for some time even after the changes in the magnet or the electric charge had stopped. Faraday admitted that he did not know what a line of force or field really was, just as Newton did not know what gravity was. He could only define these by the effects they produced. The concept of field was invented to visualise forces acting between two distant bodies but the modern view of forces gives the field a much more substantial role. It is possible to formulate mathematical equations that describe the direction and the strength of the field at every point in the space surrounding a charge or a magnet or a gravitating body. Faraday was not able to formulate these field equations but it was not long before someone else did. If Faraday had done nothing else, his invention of the concept of field would have made him famous and changed physics.

Faraday's seminal experiments on electricity and magnetism were followed through by the Scottish theoretical physicist James Clerk Maxwell (1831–1879) with an elegant theoretical demonstration (published in 1873) that electric and magnetic effects were manifestations of the same basic electromagnetic force. This was the formulation of field equations of electromagnetism. Maxwell's contribution to this subject is truly staggering. Newton's laws form the starting point for every problem in dynamics. Similarly Maxwell's equations form the starting point for every problem in electromagnetic theory. In a few elegant equations Maxwell summarised all the empirical knowledge then available on electric and magnetic phenomenon. These contained a suggestion of the existence of electromagnetic waves propagated through a vacuum at a constant speed – *a speed equal to the speed of light*. Maxwell concluded that light, which also exhibited characteristics of wave phenomena, was a form of electromagnetic effect – he had thus unified light and electromagnetism. Maxwell's theoretical analysis suggested that other forms of electromagnetic waves, with a wavelength very different from that of the visible light, must exist. In 1888 Heinrich Hertz (1857–1894) confirmed this prediction. He discovered radio waves, whose wavelength was considerably longer than the wavelength of visible light.

Maxwell's theory of classical electromagnetic radiation was of immense significance; it not only brought together the phenomena of electricity, magnetism and light in a unified framework but also made a fundamental revision of the then-accepted Newtonian way of thinking about the forces in the physical universe. The development of the theory constituted a conceptual revolution that lasted for nearly half a century. It began with the seminal work of Michael Faraday, who published his article 'Thoughts on Ray Vibrations' in May 1846, and came to fruition in 1888 when Hertz succeeded in generating electromagnetic waves at radio and microwave frequencies and measuring their properties. Maxwell's four field equations represent the peak of progress in classical electromagnetic theory. Subsequent developments in the theory have been concerned either with the interaction between electromagnetism and atoms or with the practical and theoretical consequences of the field equations. Maxwell's formulation has withstood the revolutions of relativity and quantum mechanics.

Maxwell predicted that electromagnetic disturbances travelling through empty space have electric and magnetic fields at right angles to each other and that both fields are perpendicular to the direction of propagation of the wave. He concluded that the waves move at a uniform speed equal to the speed of light. Both Maxwell and Hertz were puzzled and rather disturbed by this wave propagation of electric and magnetic fields. What was the medium enabling waves to propagate in regions free of known matter? Maxwell decided to adopt a notion that went back to antiquity – a notion that had been revived by Thomas Young (1773–1829) and Augustin-Jean Fresnel (1788–1827) in the early nineteenth century to explain the wave propagation of light. This was the concept of ether and Maxwell believed that the propagation of electromagnetic effects including light could be explained in terms of stresses in ether. It should be stressed that Maxwell's attitude to ether was highly ambivalent and he may have taken it as a working hypothesis. He did, however, believe that if ether existed it could be detected through its effect on light (see next paragraph). Maxwell's electromagnetic waves propagate at the speed of light and the magnitude of the speed was independent of 'the state of rest or motion' of the source of waves. This constancy of the velocity of light is built into the laws of electromagnetism. But, and this is the problem, 'the state of rest or motion' relative to what? To Maxwell (and to physicists before Einstein) the answer was obvious: relative to the all-pervasive ether, of course. To these physicists the ether was at absolute rest in the universe and relative to this, absolute motion could be measured. Observers at rest with respect to this ether were considered to be in a favoured position with respect to laws of nature. They were able to measure the absolute speed of light; that is, the speed of light in vacuum (i.e. a vacuum except for the presence of ether) for a source at rest with respect to the ether. For observers in motion relative to the ether the measured speed of light would depend on the speed of the observer with respect to the ether. By Newtonian relativity (or applying Galilean transformation) the speed of light measured by an observer moving with the source of light would be equal to the difference between the absolute speed of light and the speed of the observer. For an observer moving towards a source of light the measured speed of light would be the sum of the absolute speed and the observer's speed. The nemesis of this reasoning appeared in the form of the Michelson–Morley experiment.

The most celebrated experiment to detect ether was performed in 1887 – at the Chase School of Applied Sciences in Cleveland (Ohio, USA) – by Albert Michelson (1852–1931), a physicist, and his colleague

Edward Morley (1838–1923), a chemist. Michelson was a genius in instrumentation and experimentation. In 1907 he became the first American scientist to receive the Nobel Prize for physics for 'development of optical precision instruments . . .'. He had first attempted the 'ether drift' experiment in 1881 when he was doing post-graduate work at Helmholtz's laboratory in Berlin, Germany. Michelson was inspired[2] to do this experiment by remarks in Maxwell's paper (published after his death) in the scientific journal *Nature*. In this paper Maxwell conjectured that the presence of ether could be deduced by measuring the velocity of light from the eclipse of Jovian satellites. But measurements of high accuracy would be required and these were not possible in the late 1870s. Maxwell also conjectured that the experiment to detect the presence of ether was not possible on Earth. This is because to measure the velocity of light it is necessary to return the beam along its original path and on the return path the change in the speed on the outward path would be compensated. The trick Michelson employed was to compare two beams travelling at right angles to each other. Since one is concerned with measuring the change in the speed of light and not the determination of the absolute value of the speed, it is sufficient to compare two beams along tracks at right angles. He invented a new interferometer to do this experiment. The interferometer made use of the phenomenon of optical interference between two beams of light. (If two waves meet at a point and they are in phase, the result will be the sum of the amplitudes of the two waves. If the waves are out of phase, the peak of one will coincide with the trough of the other and two waves will cancel each other. Optically this would appear as alternate light and dark bands called interference bands or fringes.) The separation between the interference bands (or fringes) depends on the extent to which the two superimposed beams of light cancel or augment each other. The separation of fringes is proportional to the difference in the speed of the two interfering waves. Michelson's interferometer was capable of measuring a difference of one ten-billionth in the velocity of light. If the ether theory was correct, the time required for light to pass between two points on Earth's

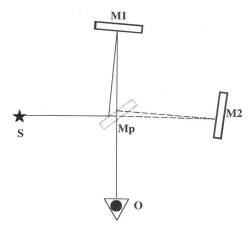

FIGURE 5.4 A schematic representation of the Michelson–Morley interferometer. A beam of light from the source S is split, at the partially reflecting mirror Mp, towards mirrors M1 and M2. The beams are reflected back and combined at the observer O.

surface should depend on the direction in which it travels with respect to the ether. The velocity of a beam of light between two points positioned west–east will be different from the velocity measured between two points positioned north–south. The measurement of this purported difference in the speed of light was within the experimental capability of Michelson's interferometer. In the first experiment, done in Germany, Michelson found no evidence of the putative ether but to his disappointment his work received little attention. In 1887 he decided to repeat his experiment, this time in the United States in collaboration with Morley from Case Reserve University. Michelson and Morley built a new interferometer along the same lines as the one Michelson had built in Berlin. He introduced a number of innovations to minimise any perturbing influences; for example, the interferometer was floated on mercury to reduce effects of vibrations. The general layout of their interferometer is shown in Figure 5.4. A beam of light (from source S) is split by a partially silvered mirror (Mp) and sent in two directions at right angles. The beams are then reflected back (at mirrors M1 and M2 respectively) and combined. The two arms were about 11 metres long. The interference fringes in the combined beam were observed at O. The change in the speed of light would be greatest if one arm of the interferometer was pointed along the direction of motion of

Earth in its orbit round the Sun. The velocity of Earth around the Sun is 30 kilometres per second, about ten thousand times less than the speed of light, but the interferometer which Michelson and Morley had built was quite capable of measuring a change of this magnitude in the speed of light. Michelson and Morley noted that the velocity of light did not change whether a beam was travelling with or against the ether. If this result had been demonstrated in the sixteenth century it would have been interpreted as a proof that Copernicus was wrong and Ptolemy right (that is, the Earth was stationary and did not move around the Sun), but by 1887 such an interpretation would have been considered incredible if not laughable. The Michelson–Morley experiment demonstrated that the speed of light was a genuine universal constant, it was independent of how it was measured. This time the scientific community did take notice of the null result of this experiment. Lord Kelvin referred to it in 1900 and noted that the null result had cast a 'cloud over the dynamic theory of light'. The full significance of the result was only realised in the early twentieth century and it stunned the scientific community.

A number of attempts were made to reconcile the observations of the Michelson–Morley experiment with the Newtonian concepts of space and time, but all appeared contrived. The majority of physicists, being by nature conservative, were confident that given time and further experiments the Michelson–Morley results would be explained with Newtonian mechanics. But there were a few who saw a deeper significance in the results. One of the first to accept the Michelson–Morley result at face value was the Irish physicist George F. FitzGerald (1851–1901). In a short article published in 1889 in the American journal *Science* he suggested that this result could be understood if the length of a material body changes as it moved through the ether. A tiny contraction along the direction of the Earth's motion (and no contraction in the direction perpendicular to the motion) would account for the null result of Michelson and Morley. FitzGerald pointed out in his article that this result would be expected from Maxwell's theory that matter was held together by electromagnetic forces.

The great Dutch physicist Hendrik Antoon Lorentz (1853–1928) also accepted the results of the Michelson–Morley experiment. Lorentz's contribution to twentieth-century physics was enormous. His central aim was to construct a single theory to explain the relationship between electricity, magnetism and light. Lorentz was appointed professor of mathematical physics at Leiden University, the Netherlands, in 1878. He had submitted his doctoral thesis just three years earlier. In his thesis he refined the electromagnetic theory of Maxwell to explain more satisfactorily the reflection and refraction of light. Maxwell's theory deals with propagation of electromagnetic waves (light and all other forms of radiation) through vacuum (and ether, if it exists). In its original form it could not account for the propagation of light through (say) glass (the speed of light through glass is not the same for all colours), or reflection off metal mirrors. According to Maxwell's theory, oscillating electric charges produce electromagnetic radiation. It was known that light was a form of electromagnetic radiation, so Lorentz proposed that there might be charged particles (electrons) in atoms whose oscillations produce light. If this were true, then a strong magnetic field would affect these oscillations and alter the wavelength of the light produced. Zeeman, a pupil of Lorentz, discovered this effect (now known as the Zeeman effect) in 1896. For this discovery Lorentz and Zeeman shared the physics Nobel Prize in 1902.

To explain the null result of the Michelson–Morley experiment Lorentz introduced (in 1895) the concept of local time, that is that the time-rate depends on the location of experiment or observation. This led him to the notion that moving bodies can contract in the direction of motion, like in the contraction proposed by FitzGerald. Lorentz derived mathematical expressions, the Lorentz transformations, to account for the null result of the Michelson–Morley experiment But for Lorentz these transformations were just mathematical constructs to explain the experimental result, they had no deeper physical significance. Between 1900 and 1904 Lorentz (and the French physicist Henri Poincaré) developed a theory of relativity in which the absolute motion of a body relative to the hypothetical ether was no longer relevant.

Poincaré named the theory the principle of relativity in a lecture given at the St Louis Exposition in September 1904. The more general formulation of the special theory of relativity and universal transformations which would self-consistently account for material events (motion of bodies) and electromagnetic phenomena was left to Einstein.

THEORY OF SPECIAL RELATIVITY

Einstein formulated his theory of special relativity in 1905. The 'special' here signifies an idealisation: in this theory Einstein only considered situations in which bodies are in uniform motion, and gravity and acceleration are absent. In the real world or universe this is impossible, but this idealisation enabled Einstein to gain physical insight into the problem he was attempting to solve. He returned to the real gravity-bound universe a few years later.

Albert Einstein (1879–1955) was born on 14 March 1879 in Ulm, Germany. His family moved to Munich when he was one year old. At the age of eight, Einstein was sent to the Luitpold Gymnasium, a rigidly disciplined school of nineteenth-century Germany with its harsh and regimented educational system. Einstein found school life both intimidating and boring. In 1894 his family moved to Milan in Italy and left him in Munich to complete his school education. He found this enforced separation depressing and six months later he withdrew from the school to rejoin his family in Italy. In October 1895 Einstein took, by special permission (he was two years under the minimum age of admission), the entrance examination for the Federal Polytechnic Academy in Zurich (now known as the Eidgenossische Technische Hochschule or ETH for short). He failed. But after another year of schooling at Aarau, in Switzerland, he enrolled at ETH in October 1896. He was the youngest of five students in the section that provided training for a degree to teach high-school physics and mathematics. Two of his fellow students were later to play a crucial role in his life – Marcel Grossmann who introduced Einstein to the geometry required to develop the theory of general relativity and Mileva Maric who became his first wife. At ETH Einstein studied under several of the world's renowned physicists and mathematicians. The Russian-born mathematician Hermann Minkowski (1864–1909) taught him mathematics. Later Minkowski's mathematics was to play a crucial role in the development of Einstein's theories but Minkowski was barely able to recall his student. Einstein did not get on well with any of his professors and his casual attitude to coursework did not help. Moreover, the physics coursework at ETH was based on 'well founded' physics of the nineteenth century; the 'new' physics including Maxwell's elegant electromagnetic theory was omitted completely from the syllabus. But it was precisely this 'new' physics which fired Einstein's enthusiasm, and he had to learn this physics from books written by physicists at other universities. Einstein graduated from ETH in 1900 with reasonably good grades, and expected to be offered a position of 'assistant' at ETH. But in this he was not successful and Einstein blamed his professors for this disappointment. He spent next two frustrating years attempting to secure a university position but without success. He finally accepted an appointment as a technical expert (third class) in the Swiss Patent Office in Bern. With this new-found security Einstein married his university sweetheart, Mileva Maric, in 1903.

Early in 1905 (his *annus mirabilis*) Einstein published a paper in the German journal *Annalen der Physik*, entitled *A New Determination of Molecular Dimensions*. This gained him a doctorate from the University of Zurich. Four more papers followed that year, all published in *Annalen*, and these were to change mankind's view of the physical world. The first of this was *On the Motion – Required by the Molecular Kinetic Theory of Heat – of Small Particles Suspended in a Stationary Liquid*. This paper provided a theory of Brownian motion of particles suspended in a liquid. In the second paper – *On a Heuristic Viewpoint Concerning the Production and Transformation of Light* – Einstein provided an explanation of the photoelectric effect (emission of electrons from some solids when illuminated by light). He postulated that light photons had both particle and wave-like characteristics. His third paper – *On the Electrodynamics of Moving Bodies* – was on the special theory of relativity. In this paper, published in June 1905,

Einstein argued that space and time were equivalent because the speed of light is finite and constant in all inertial frames. In a fourth paper, published in September 1905, Einstein showed that mass and energy were also equivalent, and this led to his celebrated equation, $E = mc^2$. At the age of 26, Einstein had demolished the fundamental basis of Newtonian mechanics, that of absolute and independent space and time.

Einstein's four papers brought him recognition from the scientific community (public acclaim, which he achieved later, was still many years away). He was now offered the university positions he had desired, first a full professorship at the Karl-Ferdinand University in Prague and then at the ETH in Zurich in 1912. In 1914 Einstein (with his family – he now had two sons) moved to Berlin at the Prussian Academy of Science where he had accepted a position which allowed him to continue his research with only occasional lecturing duties at the University of Berlin. Four years later, when the European nations were once again at war, Einstein published, in the *Annalen*, *The Foundation of the General Theory of Relativity*. The basis of this theory is that gravity is not a force but a consequence of curvature in space-time. This curvature is created by the presence of matter or energy.

Einstein's special theory of relativity is based on his deeply held belief in the universality of all laws of nature: if the Newtonian–Galilean principle of relativity is applicable to mechanical phenomena (moving bodies) then it should be applicable to optical (electromagnetic) phenomenon. If this were not true then optical laws would hold a special place in nature. Einstein appears to have started thinking about this problem as early as 1899. In August of that year he wrote to his sweetheart Mileva Maric[3]

> I am more and more convinced that the electrodynamics of moving bodies, as presented today, is not correct.

The problem that was exercising Einstein's mind can be visualised as follows. Suppose an observer was to chase after a beam of light and catch up with it, how would the light waves appear to this observer? Intuitively or guided by Newton's laws of motion, the observer would expect the light waves to appear stationary, that is light standing still! But this is not allowed by Maxwell's theory or by observations. This line of reasoning led Einstein to challenge the 'common-sense' notions of space and time. Einstein was uneasy that the Galilean transformations and the Newtonian principle of relativity did not apply to light. This had been demonstrated by the Michelson–Morley experiment, although it is not known if Einstein was aware of the experiment[4]. In a letter written in 1954, a year before his death, he denied that he was aware of this experiment when he formulated the special theory. But this seems unlikely. Einstein tended to work in isolation but he was not completely isolated. He was aware of the work by Lorentz, whom he admired greatly. Lorentz certainly knew of the Michelson–Morley experiment and had mentioned its implications and its importance in his writings. It should, however, be emphasised that the special theory has a far deeper purpose than an explanation of the Michelson–Morley experiment. Einstein found the concepts of absolute space and time, inherent in Newtonian physics, aesthetically unsatisfactory. Space and time had puzzled scientists (and philosophers) long before Einstein. Copernicus asserted that time was not absolute (not part of a divine design) but a product of the Earth's rotation and orbit round the Sun. Bishop George Berkeley, a philosopher adversary of Newton, dismissed absolute time, space and motion as fictions of the mind. He argued that all motion, both uniform and nonuniform, was relative to the distant stars. Leibniz, the mathematician adversary of Newton, rejected absolute space, as it was not observable and could have no observable effect.

In the theory of special relativity Einstein rejected the concepts of absolute space and time. He based his theory on just two assumptions:

- All physical laws are identical in inertial frames or all inertial frames are equivalent.
- The speed of light (electromagnetic radiation, to be precise) in vacuum has the same value in all inertial frames.

The first assumption is a generalisation of a wide range of experiences while the second states an experimental fact. Einstein's principle of special relativity can now be stated thus: no experiment or observation (mechanical or optical: electromagnetic, to be accurate) made in an inertial frame can be used to determine the state of rest or uniform translation of the experimenter or the observer. The concept of absolute motion is thus removed from the physical sciences. A different way of stating the principle of special relativity is this: the laws of nature are invariant for all inertial observers. The principle of invariance has become a powerful tool because it imposes strict conditions on laws of nature (see *Conserved quantities* in Chapter 4, Newton). A statement (or an equation) can be a law of nature if and only if it can retain the same (algebraic) form when transformed from one inertial frame to another. Einstein now turned the principle of invariance on its head by noting that since the speed of light is invariant in all inertial frames, the constancy of the speed of light is a law of nature. As the Galilean transformations do not maintain the invariant value of the speed of light, they must be replaced by transformations which give the same speed of light for all inertial observers. Einstein replaced the concepts of absolute space and time with his own concepts of relative space and relative time, which apply to all events in nature whether mechanical or electromagnetic. He replaced the old Galilean transformations with those now called the Einstein–Lorentz transformations. These transformations took into account the fundamental requirement that the velocity of light should be invariant between inertial frames. The name of Lorentz is associated with these transformations because, as described above, he had first obtained these relations to explain the results of the Michelson–Morley experiment. Einstein derived these transformations independently and from general considerations and showed that these transformations applied to both mechanical and electromagnetic events.

Let us re-consider the two frames shown in Figure 4.1. In Newtonian relativity the spatial separation (length or distance) is invariant between two frames in relative uniform motion. In special relativity the separa-

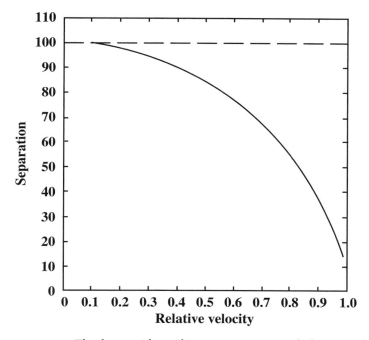

FIGURE 5.5 The change with speed in separation or interval when viewed from a frame in relative motion. The Galilean value of the separation or interval is shown by the dashed line. The relative velocity is given in terms of the speed of light. As the speed increases the separation decreases (Lorentz–FitzGerald contraction) or the interval decreases (time dilation).

tion in frame S as measured by an observer in frame S' which is moving relative to S is as shown in Figure 5.5. A length of (say) 100 metres will appear smaller as the relative speed of the frames increases; this is the FitzGerald contraction proposed to explain the Michelson–Morley experiment. Similarly a temporal separation (interval) in S as measured in S' is as shown in Figure 5.5. An interval of 100 seconds will appear as 43.0 seconds to an observer moving at 90% of the speed of light, so the clock will appear to slow down.

The Newtonian concepts of absolute space and time were based on an implicit assumption that the speed of light was infinite. The logical basis for this assumption had disappeared in 1776 when Roemer showed that the speed of light was finite. The Einstein–Lorentz transformations reduce to the Galilean form if the velocity of light is set to infinity. The transformations also reduce to the Galilean form if the relative velocity of the two frames is considerably smaller than the speed of light. In these transformations the speed of light, measured from frames in relative motion, will have exactly the same value regardless of the magnitude (or direction) of the relative speed of the frames. The Galilean transformations are thus an approximation of the more general Einstein–Lorentz transformations. Speeds encountered in everyday experience are considerably lower than a tenth of the speed of light and as seen from Figure 5.5 there is effectively no difference in the spatial or temporal intervals when these are measured from frames moving at this low speed. The core conclusion of the theory of special relativity is that *the values of spatial and time intervals measured depend on the relative speed of the inertial frames from which these measurements are made*. This ambiguity in the measured length and time arises because the concept of Newtonian simultaneity is not valid when the relative speed of frames approaches the speed of light. In Newtonian relativity, observers in frames S and S' *always* agree everywhere about simultaneity, i.e. $t = t'$ everywhere. However, in special relativity observers in frames S and S' can agree on simultaneity at one point in space-time, but they will *disagree at all other points*.

The great French mathematical physicist and philosopher Jules-Henri Poincaré (1854–1912) had questioned the objective meaning of simultaneity a few years before Einstein's special relativity. Poincaré came from a distinguished family: his first cousin, Raymond Poincaré, was the president of the French Republic during the First World War. Henri Poincaré was a sickly child and his early education was by his gifted mother. At an early age he excelled in written composition, a ability he retained throughout his life. During adolescence he became interested in mathematics and easily won top honours in mathematics during his university education. He joined the University of Paris in 1881 and stayed there for the rest of his life. He lectured and wrote extensively on physics, theoretical astronomy, relativity and philosophy. He was also a gifted interpreter of science and used his superb literary gifts to describe for the general public the meaning and importance of science and mathematics.

In a paper published in 1898 Poincaré noted that the qualitative problem of simultaneity couldn't be separated from the quantitative problem of measuring time. In the following years he developed most of the concepts of special relativity. He even *conjectured* that the velocity of light might be the upper limit of attainable velocity. In 1906, in a paper on the dynamics of the electron he obtained many of the results of special relativity. Poincaré had also realised that Newton's law of gravitation was not invariant under Lorentz transformation and therefore the law needed modification. He went on to postulate gravitational waves that propagate with the speed of light. But these were proposals and conjectures; Poincaré did not formulate a consistent theory encompassing his ideas.

It is remarkable that in his extensive writings Poincaré never mentioned Einstein, referring to him only once, in a talk given shortly before his death. In his writings he did not link Einstein and special relativity. Poincaré, one of the greatest theoretical physicists of the late nineteenth and early twentieth century, either failed to appreciate or chose to ignore special relativity and Einstein's seminal contribution to it. Einstein in turn did not mention Poincaré and his early contribution to the development of special relativity. Einstein acknowledged Poincaré's contribution only once: in 1953 Einstein was invited to the celebration of the fiftieth anniversary of special relativity which was to be held at Bern in Switzerland. He declined owing to ill health but wrote to the organisers of the conference that Lorentz and Poincaré should be suitably honoured on this occasion. This was 40 years after Poincaré's death!

From antiquity to the late nineteenth century space and time had been considered to be absolute and independent parameters of nature. Einstein showed that this assumption was at best an approximation and at worst a mistake. This does not mean that there are no 'absolutes' in nature. In 1908 Hermann Minkowski (who in 1902 had moved from ETH in Zurich to Göttingen in Germany) showed that the Einstein–Lorentz transformations have a simple geometric interpretation that is both beautiful and useful. He showed that it was possible to treat time as if it was analogous to a spatial dimension. The motion of a body can be regarded as forming a curve made of 'events', which can be identified by three spatial coordinates and a forth coordinate, time. This four-dimensional space is called Minkowski space-time and the curve is called a world-line. Space-time serves as the immutable backdrop of all physical processes, without being affected by them. Newton's First Law can be interpreted in four-dimensional space-time as a statement that world-lines of bodies not influenced by external forces are straight lines in space-time. Minkowski defined a *space-time interval* as

$$s^2 = x^2 + y^2 + z^2 - c^2 t^2$$

and he showed that this interval was invariant, that is, this interval is same for any observer in relative uniform motion. This should be compared with the invariant distance in Newtonian mechanics (see Chapter 4, Newton). The Minkowski space-time is a rigidly flat continuum, similar to the three-dimensional space of Euclid's geometry. Distances between events or world-points are measured by the invariant space-time intervals, whose magnitudes do not depend on the particular coordinate system or frame of reference used. The Minkowski space is also homogeneous: that is, a figure can be transformed between frames without distortion. Minkowski's geometric ideas provide a powerful tool for checking the mathematical consistency of special relativity and for calculating its experimental consequences. They also provide a revolutionary interpretation of motion of a body: all bodies are constantly moving through space-time, along all four dimensions – the three dimensions of space and one of time. A 'stationary' body is at rest in the three spatial dimensions but is moving in the fourth time dimension. If this body moves spatially it derives its spatial movement from the motion in the time dimension – the 'sum' of the motion in the four dimensions always stays constant. Thus a spatially moving body slows down in the time dimension or its clock will slow down. This is the time dilation described earlier. Einstein was not impressed by Minkowski's discovery; he felt that the mathematical rigour obscured the physical ideas enshrined in special relativity. Four years later, in 1912, Einstein had to 'eat humble pie' as he came to realise that Minkowski's absolute space-time was essential for formulating the general theory of relativity. This is because Minkowski's ideas have a natural generalisation, which incorporates the effects of gravity.

It is worth comparing the invariant space-time interval with the expression of distance measured in three-dimensional space, that is a general definition of Pythagoras's theorem (as described in Chapter 2, Kepler):

$$s^2 = x^2 + y^2 + z^2$$

The similarity of these two expressions demonstrates that the Minkowski space-time interval is really an expression of the Pythagoras theorem in four-dimensional Euclidean geometry. Pythagoras's theorem gives the distance between two points, whereas Minkowski's space-time interval gives the distance between two *events*, one at the point (x_1, y_1, z_1) at time t_1 and the other at point (x_2, y_2, z_2) at time t_2. The geometry of special relativity is Euclidean and is applicable only to inertial frames (that is, frames in uniform relative motion) and not to accelerated frames. It is, however, worth remembering that although in special relativity space and time are treated as four identical coordinates, the three-dimensional space and the one-dimensional time are physically quite distinct. The space dimensions can have both positive and negative values but the time dimension can only be uni-directional, a negative time is physically meaningless.

Newtonian mechanics leads to three laws of conservation – linear momentum, energy and angular momentum (see Chapter 4, Newton). Newton's Second Law also implies conservation of mass. The conservation of linear momentum, mass and energy are incompatible with Einstein's theory of special relativity and Einstein showed that these three quantities are related through

$$p^2 - (E^2/c^2) = m_0^2 c^2$$

where p is the linear momentum, E is the *total* energy of the body and m_0 is the 'rest mass' of a body. Einstein defined the rest mass (or the

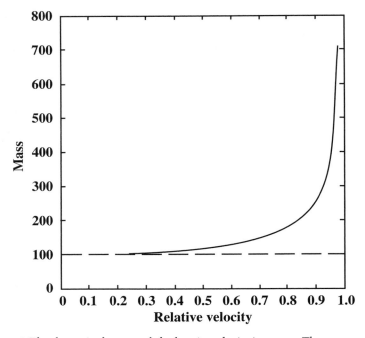

FIGURE 5.6 The change in the mass of a body as its velocity increases. The non-relativistic mass is shown by the dashed line. The speed of the body is given in terms of the speed of light.

proper mass) as the mass of a body when the body is at rest with respect to an inertial observer. He showed that it is necessary to differentiate between the rest mass m_0 and mass m when a body is in motion relative to an observer. To an observer at rest the mass of a body in motion will appear to increase as the speed of the body increases, as shown in Figure 5.6. If a body travels at the speed of light its mass becomes infinite, which is physically meaningless. To put it differently, information cannot be transmitted at speeds equal to or higher than the speed of light. On the other hand, bodies of zero rest mass (like light photons) can only travel at the speed of light. Notice that the increase in mass is relative to an observer at rest with respect to the body. If the observer is moving with the body then with respect to this observer the body is stationary and there is no change in its mass. The mass increase does not have to be considered at 'everyday speeds' as these are considerably lower than the speed of light.

The increase in mass with speed has now been confirmed experimentally. This increase has to be taken into account in designing particle accelerators, in which speeds close to the speed of light are attained. For example, in an electron synchrotron, used to accelerate electrons to high speeds, the relativistic mass of electrons can equal 10 000 times their rest mass.

The total energy of a freely moving body can be stated as

$$E^2 = c^2 p^2 + m_0^2 c^4$$

The total energy depends on the linear momentum (which changes from observer to observer) and on the rest mass (which is the same for all observers)[5]. It is interesting to note that in Newtonian mechanics the total energy of a body (which is equal to $p^2/2m_0$) is zero if the motion (or the linear momentum) of the body is zero. In the special theory of relativity, if a body is at rest or if its linear momentum is zero, the energy of the body is

$$E = m_0 c^2$$

This is the most celebrated equation in modern physics. In the special theory Einstein replaced the Newtonian laws of conservation of inertial mass, momentum and energy with a single law of conservation, that of *total* energy. This equation also expresses the equivalence between mass and energy, without which it would be impossible to understand nuclear weapons and the vast luminosity of the stars.

Newtonian mechanics has three absolute quantities, space, time and mass. The nineteenth-century scientists introduced a fourth absolute quantity to physics, namely energy. Einstein showed that these four quantities were related and reduced the four quantities to two pairs – space-time and mass-energy – and showed that the pair members were related by the speed of light.

Time dilation as predicted by special relativity has now been confirmed experimentally. Elementary particles called muons are formed at about 10 kilometres in the Earth's atmosphere by collisions of incoming cosmic rays with atoms of the gas in the upper atmosphere. The lifetime of this particle is about 2×10^{-6} seconds, and at the speed of light the muons should decay in about 600 metres and should not reach the surface of Earth. But at speeds close to the speed of light the 'clock of the particle' slows down and its lifetime increases by over a factor of 200 relative to an observer on Earth, so the particles easily travel a distance of 10 km, to be detected on Earth. To the muon however it is the Earth that appears to be moving at speeds close to the speed of light; the 10 km distance appears considerably shorter and is easily traversed. In a experiment done in 1966 at the European high-energy laboratory near Geneva, Switzerland (CERN), muons were deflected by magnetic fields to move in circular paths, and so could be stored. The speed of these 'stored' particles was about 99.7% of the speed of light and the observed increase in their lifetime agreed with predictions of special relativity with 2% accuracy.

The strength and the predictive power of special relativity were demonstrated in 1928 when the British physicist Paul A. M. Dirac merged this theory with the theory of quantum mechanics. This merger suggested that all elementary particles have a property called 'spin' and that antimatter should exist. Both have now been confirmed experimentally. The merging also produced a self-consistent picture of the atomic structure leading to a deeper understanding of the atomic shells that govern chemical reactions. In the 1940s two American physicists, Richard Feynman and Julian Schwinger, and the Japanese physicist Shinichiro Tomonaga succeeded in simultaneously merging special relativity, electromagnetism and quantum mechanics and thus provided a rational explanation of relativistic phenomenon in the subatomic realm. This was the formulation of one of the most important and successful theories of the twentieth century – the theory of quantum electrodynamics (QED). The stunning experimental demonstration of this theory of fundamental particles and forces is also an endorsement of the theory of special relativity.

THEORY OF GENERAL RELATIVITY

Einstein developed his theory of special relativity out of a deep conviction that in nature there were no preferred frames of reference. In this theory the concepts of inertial frame and Euclidean geometry are retained but the Galilean transformations are abandoned. But it was clear to Einstein that the special theory does not remove all restrictions on coordinate systems since it says nothing about the validity of laws of nature in noninertial frames of reference (e.g. accelerated frames, rotating frames or frames in a gravitational field). Newton's law of gravitational attraction is not the same between inertial frames moving at high relative speed (Poincaré had recognised this problem, as described above). This is because the distance and inertial mass, parameters used to express gravitational attraction between two bodies, have no absolute values but depend on the velocity of the observer and the bodies. The gravitational attraction between two approaching bodies, measured by an observer on one of the bodies, will be higher than the attraction between the bodies when the bodies are at relative rest. The increase in the mass and the decrease in the separation of the approaching body cause this difference in the gravitational attraction. Thus Newton's law of gravitational attraction does not conform to the invariance requirements of the special theory. The instantaneous action at a distance required by Newtonian gravity is also contrary to the basic principle of special theory which asserts that a physical effect cannot be transmitted at speeds greater than the speed of light.

To eliminate the flaws inherent in the special theory Einstein introduced the principle of covariance:

- The laws of nature have the same mathematical form in *all* frames of reference.

Compare this with his earlier principle – a law of nature must have the same form in *all inertial* frames – which he introduced in special relativity. By introducing the principle of covariance, Einstein insisted that all motions, uniform or accelerated, were relative. A relative accelerated motion appears contrary to our everyday experience. If we were in a smoothly flying plane we would be unaware of our motion, but if the plane banked or altered its speed we would immediately notice the accelerated motion. This would lead us to believe that accelerated motion was indeed absolute and different from uniform motion. But consider an astronaut standing on weighing scales in a rocket. When the rocket is launched it will accelerate away from Earth and the astronaut's weight will appear to increase because of the acceleration of the rocket. If now the rocket motor is switched off the rocket will begin to fall, the astronaut will float freely in the rocket and in that frame the astronaut will appear to be weightless. The astronaut would be led to believe that there was no force acting on him or her. But to an observer at the launch pad the action of gravity is quite obvious. Such 'thought experiments' led Einstein to formulate his fundamental *principle of equivalence*.

- There is no way for an observer in a non-inertial frame (i.e. accelerated frame) to distinguish between gravitational force and inertial forces acting on bodies in that frame.

This principle does not just assert that gravitational and inertial forces are equivalent but insists that every effect produced (on any physical system) by acceleration, or observed by an observer in an accelerated frame can be reproduced by an appropriate gravitational force and observed by an observer at rest in the gravitational field. To put it another way, the equivalence principle prevents an observer from detecting uniformly accelerated motion. Observed accelerated motion could be attributed either to acceleration in gravity-free space or to a gravitational field. Compare this with the principle of relativity – this principle prevents an observer from detecting uniform motion (see *Theory of special relativity*, earlier in this chapter). The equivalence principle follows naturally from Galileo's so-called 'Leaning Tower of Pisa' experiment demonstrating the equality of gravitational and inertial mass. Galileo had shown that all bodies fall down to Earth with the same acceleration (neglecting the resistance due to air); that is, the acceleration of a body is independent of its mass or composition. From a relativistic point of view, it is entirely consistent to consider that it is the ground that has accelerated to the bodies, which are at rest. In this case the composition of the bodies is irrelevant, and all bodies should appear to be approached by the ground at the same rate of acceleration. Newton had tested the equality of gravitational and inertial mass with his pendulum experiments. In the late nineteenth century the Hungarian scientist Baron Eötvös performed precise experiments to show that the inertial and the gravitational mass were equal to an accuracy of one part in a billion (see Chapter 6, Dicke). Einstein apparently did not know of these experiments but referred to them extensively after the work had been pointed out to him. Newton accepted the equality of inertial and the gravitational mass as a phenomenological fact and ascribed it no special importance, but to Einstein this equality was not a mere accident of nature, he saw a profound physical significance in it. The principle of equivalence forms a cornerstone of general relativity and its verification is crucially important. Increasingly precise experiments have been performed to test it and the current limit is about one part in 10^{12} (see Chapter 6, Dicke).

Einstein began to formulate the general theory sometime in 1906 and it was to be eight long years before he was ready to present it in its full intellectual rigour. During those eight years Einstein moved between a number of universities in Europe but 1912 appears to have been crucial to the mathematical development of the theory. In August that year Einstein returned from Prague to Zurich. By now he had convinced himself that time and light were affected by gravitation, but these ideas had to be put on a firm mathematical basis. In Zurich Einstein turned to his friend and former fellow student Marcel Grossmann for help. Grossmann was then a professor of geometry and the dean of the mathematics and physics section of ETH. He pointed out to Einstein that to solve the problem of gravitation he would need a space-time possessing the Riemannian geometry as opposed to the flat Euclidean geometry of special relativity. Einstein was blissfully unaware of Riemann and of his multi-dimensional geometry and its significance to his work. The transition from Euclidean geometry to

Riemannian geometry was the crucial step that led Einstein, initially with Grossmann's collaboration, to his ultimate formulation of post-Newtonian gravity. In 1914 Einstein moved to Berlin and on 25 November 1915 he presented to the Prussian Academy of Science his paper *The Field Equations of Gravitation*.

The physical consequences of the gravitational field in the general theory of relativity formulated by Einstein can be summarised as follows: space-time is a four-dimensional non-Euclidean continuum, the curvature (or warping) of the continuum being a consequence of the local distribution of matter or energy. Particles and light rays travel along the geodesic (stationary distance) of this four-dimensional geometric world.

In the general theory, space-time is a four-dimensional continuum as in the special theory. But this is where the similarity ends. In the general theory the invariant intervals are defined *only locally* between events taking place *close to each other*. Only small regions of space-time resemble the continuum envisaged by Minkowski, just as small sections of a sphere appear nearly planar. Far from being rigid and homogeneous, the general-relativistic space-time continuum has geometric properties that vary from point to point and are affected by local mass or energy. General relativity thus makes geometry part of physics and properties (such as curvature) of the space-time defined by this geometry can be studied by means of scientific experiments. The basic idea in the theory of general relativity has been summarised thus: space-time tells mass how to move, and mass (or energy) tells space-time how to curve. There are two principal consequences of the geometric nature of gravitation: (1) the acceleration of bodies depends only on their mass and not on their chemical or nuclear constitution, and (2) the path of a body or light rays in the vicinity of a massive body is different from that predicted by Newtonian mechanics.

Just five days before Einstein presented his paper to the Prussian Academy of Science, David Hilbert (1862–1943), one of the greatest mathematicians of all time, presented to the Royal Academy of Science in Göttingen the mathematical framework on which the general theory of relativity is based. Hilbert had followed Einstein's work and was fascinated by his ideas. In the summer of 1915 he invited Einstein to Göttingen to give seminars on the work he was doing. In the following months he pondered over what he had learned from Einstein's summer seminars. In the autumn, during a vacation, the key ideas fell into place and within few weeks he had formulated the elegant form of Einstein's field equations, which describe how mass curves space-time[6]. Einstein had arrived at the same result after a number of diversions down blind alleys and after months of frustrating trial-and-error. Einstein and Hilbert fell out over this because Einstein felt that Hilbert had stolen his thunder, but they made up after a few months and no lasting rift was produced. The credit for the theory of general relativity is, rightfully, given to Einstein. He had the physical insight such as the equivalence principle and the geometric form of the theory; the 'paradigm shift' was entirely due to Einstein. Hilbert had taken the last mathematical step, and although his was an intellectually elegant step it was, nevertheless, only the last step. Einstein (and Hilbert) had formulated the field equation of gravitation, a goal that had eluded Riemann 50 years earlier. Einstein's field equation does not describe the magnitude and direction of force in the vicinity of a gravitating body, as Maxwell's field equations do for a charged body. Rather, it describes the curvature of space-time in the vicinity of a body.

In the theory of general relativity Einstein reasoned that gravity was linked to space-time and the linking agency was Riemannian geometry. To illustrate this, consider the motion of two bodies, one on a perfectly flat surface and the other on the surface of a perfect sphere. The body on the flat surface will continue to move in a straight line but the body on the sphere will move along a curve determined by the surface of the sphere. According to Newton's First Law, there is no force acting on the body on the flat surface but the body on the sphere is constrained to move along the curve by a force that is always directly towards the centre of the sphere. This was Newton's explanation for the near-circular orbits of planets, the inward directed force being gravity. Einstein asserted that this was an illusion, there was no force constraining the body on the sphere to move along a curve but *the gravitational field was a distortion of geometry of space-time from Euclidean to non-Euclidean form*. If we accept this then the motion of a body on a flat surface and a spherical surface are equivalent. It is now necessary to restate Newton's First Law – a free body moving in any frame of reference moves along a path that is the stationary distance between any two points on the path. In Euclidean geometry the 'stationary distance between two points' is a straight line but in the non-Euclidean geometry this distance is a geodesic. This restatement of Newton's First Law may appear to contradict our everyday experience. If an object is thrown at any angle to the vertical it moves along an arc of a parabola which is certainly not the shortest geometrical distance between the point from which the object started and the point where it comes to rest. The shortest geometrical distance between these two points is a straight line. Why does the body not move along this straight line? But note that the straight line is in a three-dimensional space. The path of a body in motion has to be considered in a four-dimensional space-time and not just in a three-dimensional space. The parabola is the three-dimensional projection of the four-dimensional stationary path – a geodesic.

The special theory of relativity synthesised the separate Newtonian concepts of three-dimensional space and one-dimensional time into a single four-dimensional Euclidean space-time continuum. The general theory retains the four-dimensional space-time continuum as the geometrical framework in which the laws of nature are to be stated, but the continuum is non-Euclidean. In general relativity the curvature of the space-time is determined by the local distribution of matter (or energy): the greater the density of matter in a region, the higher the curvature of the space-time in that region. It is worth noting that in his general relativity Einstein was not attempting to find a different interpretation of Newtonian gravity or to 'fix' special relativity to include gravity, he was attempting an entirely new interpretation of gravity. In the special theory of relativity space and time are combined, but space and time are still a fixed background in which events happen. It is possible to choose different paths through space-time, but the background

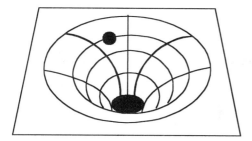

FIGURE 5.7 Distorted space-time and a 'new' interpretation of fall of an object to Earth. The space-time is distorted by mass or energy, thus (the mass of) the Earth distorts the space-time around it. A 'falling' object rolls along the geodesic of curved space-time near the Earth.

of space-time is not modified. In general relativity, gravity is no longer a (Newtonian) universal force that operates in a fixed background of space-time. Instead it is a property (curvature or distortion) of space-time caused by local mass or energy. It is interesting to note that since gravity is a universal force there can *never* be a truly inertial frame and special relativity will *always* be an approximation. An important property of mass and energy is that they are always positive and that is why gravity is always an attractive force. Repulsive gravity has never been detected. According to general relativity, this means that space-time is curved back on itself like the surface of a sphere or the curvature of space-time is always positive. If mass (or energy) had been negative, or gravity repulsive, space-time would have had a negative curvature like the surface of a saddle. When Einstein applied general relativity to explain the universe, he regarded the positive curvature of space-time as a problem (see Chapter 7, Hubble & Eddington).

Our everyday experiences now have to be reinterpreted in the context of Einstein's general relativity. For example, an apple falling to Earth is no longer considered to be attracted by some mysterious force acting at a distance through space but instead rolls into the local space-time 'well' created by Earth (Figure 5.7). The simplest way to visualise this statement is to consider space-time as a sheet of rubber stretched flat. A heavy object distorts the sheet by stretching it locally, the amount of distortion depending on the mass of the object. The Sun, being the most massive object in the solar system, causes the largest distortion of the space-time in its immediate vicinity. That curvature curves space further out, and so on. The planets are trapped in this well surrounding the Sun.

Using the principle of equivalence, Einstein deduced the advance of perihelion of Mercury. This provided a splendid test for his theory. The orbit of Mercury is not fixed but precesses and its perihelion advances, as described earlier. The anomalous advance of 43 arcseconds per century had baffled astronomers for almost 50 years before Einstein. The answer lies in the interpretation of Kepler's Third Law within the context of general relativity. The Third Law states that the square of the period of a planet is proportional to the cube of the planet's mean distance from the Sun. These two quantities are measured in two very different frames of reference. The period is measured with a clock on Earth and is not affected by the Sun's gravity at the orbit of Mercury. But the mean distance can only be measured with a measuring rod placed near Mercury, in the local gravitational field. As Mercury moves closer to the Sun the measuring rod shortens in the gravitational field and therefore the measured distance is larger and by Kepler's Third Law the period will be larger. On successive orbits Mercury will move faster in its orbit as it approaches the Sun and the perihelion will advance. Einstein calculated an advance of 42.98 arcseconds per century, in excellent agreement with the observed value. This was the first of three classical tests of general relativity.

Einstein calculated the precession of Mercury's orbit in late 1915 as he was putting the finishing touches to his theory of general relativity. This was the first observable quantity calculated with the new theory and Einstein was beside himself with joy at the close agreement. The problem of Mercury's orbit appeared to have been solved, but it raised its head again in the 1960s. The method of accurately determining the orbits of planets by radar ranging (see Chapter 6, Dicke) was developed in the 1950s and these measurements suggested that the perihelion shift of Mercury's orbit was indeed 42.98 arcseconds per century within an accuracy of 10%. The agreement with general relativity was staggering. But in 1967 the American physicist Robert Dicke (more about him in Chapter 6) raised the question of the 'figure of the Sun'. Newtonian theory suggests that a rotating spherical mass should have an oblate spheroidal shape, that is, flattened at the poles. For the Sun the expected flattening is quite small: the polar diameter should be only about 200 metres smaller than the equatorial diameter. However, Dicke (and his collaborator H. Mark Goldberg) measured a flattening of 52 kilometres. This flattened shape of the Sun could account for about 3 arcseconds per century in the advance of Mercury's perihelion and general relativity would have to account for an advance of about 40 arcseconds per century. This is significantly different from the precise value of 42.98 arcseconds per century calculated by Einstein. The perihelion advance of Mercury calculated with the Dicke–Goldberg oblateness of the Sun seemed to cast doubt on Einstein's general relativity. Interestingly this revised value of the advance appeared to be in agreement with the Brans–Dicke theory of gravity – one of the many alternative theories of gravity (discussed later in this chapter) that were developed in the 1950s and 1960s. The argument about the 'figure of the Sun' raged for almost 20 years and was partly responsible for triggering renewed interest in general relativity and experimental verification of the theory (see Chapter 6, Dicke). But more precise measurements of Sun's oblateness (the most recent with the solar observatory SOHO) suggest that the polar flattening is really as small as that suggested by the Newtonian theory. Also, the Brans–Dicke theory has been found to disagree with a number of experimental tests and at present Einstein's general relativity rules supreme.

The equivalence principle suggests that a photon loses energy as it rises through a gravitational potential (*gravitational redshift*) or, what comes to the same thing, clocks run slower in a gravitational field (*gravitational time dilation*). Einstein had predicted this eight years before the full formulation of general relativity. He returned to this in his seminal paper *The Foundation of the General Theory of Relativity* published in 1916. In this paper he states:

> Thus the clock goes more slowly if set up in the neighbourhood of ponderable masses. From this it follows that the spectral lines of light reaching us from the surface of large stars must appear shifted towards the red.

Einstein regarded this as a fundamental prediction of his general theory but it is now recognised as a test of the equivalence principle. Any theory of gravity that is compatible with the equivalence principle will predict the same gravitational redshift, and in this sense the redshift is really a consequence of the curved space-time.

It is interesting to note that Newtonian mechanics also predicts redshift of photons escaping a gravitational field. The reasoning behind this is that light is a form of energy and hence from Einstein's mass–energy relation, it has an equivalent mass. This mass must expend energy as it escapes from a gravitational field. For a photon the loss of energy is equivalent to moving towards the red end of the spectrum or a redshift. Newtonian redshift is due to a change in the mass of the photon. Gravitational redshift was another test of Einstein's general relativity. But this measurement is not easy; astronomical observations are not appropriate because it is difficult to disentangle the gravitational redshift from the Doppler shift caused by local velocities in the atmosphere of a star. Instrumentation of high sensitivity and spectral resolution is required to measure the redshift in laboratory experiments. The laboratory confirmation of gravitational redshift was only possible in the 1960s with precision techniques developed for nuclear physics (see Chapter 6, Dicke). Nowadays the global positioning satellites (GPS) routinely measure gravitational time dilation.

Einstein also used the equivalence principle to predict that a beam of light would bend in a gravitational field. To illustrate this, consider a reference frame in a gravitational field. To an observer at rest with a reference frame (Figure 5.8(a)) a beam of light will appear to move in a straight line between two points. But to an observer at rest with the gravitational field, the beam will appear to bend (Figure 5.8(b)). In both situations the box (the reference frame) is falling in the gravitational field and the beam strikes the same spot on the 'box'. But in Figure 5.8(a) the observer moves with the box (that is, at rest with the reference frame) and in Figure 5.8(b) the observer is remote from the box. A remote observer will thus see a beam of light bend in a gravitational field. Einstein had first calculated the angle of deflection using

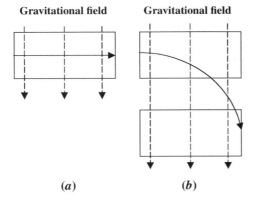

FIGURE 5.8 In situation (a) the observer is at rest with the frame of reference and the beam of light appears to move in a straight line relative to the observer. In situation (b) the observer is remote, that is, at rest with the gravitational field, and the beam of light appears to bend relative to the observer.

the (Newtonian) argument similar to that given for gravitational redshift – that is, light photons have effective mass and should therefore feel the gravitational pull of a massive body.

Almost 100 years earlier, in 1783, Reverend John Michell had considered the influence of gravity on light. Newton's corpuscular theory of light was in vogue at the time and Michell reasoned that light would be attracted in a gravitational field in the same way as ordinary matter. Fifteen years later the great French mathematician Pierre Laplace came to the same conclusion. Prompted by this work the German astronomer Johann George von Soldner (1776–1833) determined the angle through which a light ray would be bent if it skimmed the surface of the Sun. Soldner was a self-taught man who became a highly respected astronomer. He made fundamental contributions to the field of astrometry (precision measurement of astronomical positions) and became the observatory director at the Munich Academy of Science. In 1803 Soldner showed that a ray of light grazing the limb of the Sun would be deflected by 0.875 arcseconds. This work was published in a German astronomical journal and was promptly forgotten. This was for two reasons: first, measurement of the calculated deflection was beyond the technical capability of astronomical instrumentation at the beginning of the nineteenth century and second, the corpuscular theory of light was going out of fashion and giving way to Huygens' wave theory of light. According to the wave theory, light moves through the ether and is therefore not deflected in a gravitational field. In 1907 Einstein recognised that the equivalence principle implied that a gravitational field would influence a beam of light. In 1911, he computed that the beam of light grazing the Sun would be deflected by 0.875 arcseconds. He proposed that this effect should be looked for during a total solar eclipse. During an eclipse the stars near the Sun would be visible and would appear displaced from their normal position. But a deflection of 0.875 arcseconds was still close to the limit of 'detectability' of the available astrometric techniques and Einstein's proposal was forgotten. This was singularly fortunate, as will be shown in a little while.

By 1915 Einstein had the complete theory of general relativity and he realised that light grazing the Sun would move along the geodesic of space-time curved by the Sun's mass and the deflection of a light beam would be larger than that given by just the equivalence principle. The two deflections are almost equal, so the net deflection is 1.75 arcseconds. This deflection, although small, is measurable and the measurement was undertaken by the great British astrophysicist Arthur Eddington (see Chapter 6, Dicke). He learned about Einstein's work from a colleague, Willem de Sitter, in the neutral Netherlands (in 1915 Britain and Germany were at war) and realised the importance of this 'prediction'. He started preparations to make the necessary observations during the total solar eclipse expected in 1919. Eddington was a Quaker and a conscientious objector and the British Military Tribunal had granted him exemption from active service. The argument that weighed heavily with the tribunal was Eddington's involvement in the preparations for the forthcoming eclipse expedition to test Einstein's theory of general relativity. The results of the measurement were in agreement with the value calculated by Einstein and not with that calculated according to Newton. The result was announced by the Astronomer Royal, Sir Frank Dyson, to a crowded meeting of the Royal Society in London. The announcement caused a sensation and overnight Einstein became a household name.

Suppose it had been possible in 1911 to measure the deflection of stars due to the gravitational field of the Sun and a successful expedition had been mounted. This expedition would have measured a deflection twice as large as that predicted by Einstein at the time. The new prediction of Einstein in 1915 would then have appeared as a modification made to bring the theory in agreement with observations and general relativity would not have had the impact it had. A lingering doubt would have persisted that the theory had been 'fixed' to agree with observations. As it is the results of 1919 expedition provided a clear choice between the Newtonian theory (and Einstein's partial theory of 1911) and the complete theory of general relativity. These results proved unambiguously that gravity had to be interpreted along the lines defined by Einstein's general relativity and not according to Newton's laws of gravitation.

In 1921 Einstein was awarded the Nobel Prize for his explanation of photoelectric effect and 'work in the field of theoretical physics'. Eight years later he published the first version of his unified field theory. Although the paper was well received, the preliminary nature of the theory was apparent. Undaunted, Einstein continued working on the theory. But more disturbing and dangerous events were to intervene in his life. Rising militarism and anti-Semitism in Germany forced Einstein, in 1933, to renounce his German citizenship and leave Germany. He accepted a full-time position as a founding member of the new Institute of Advanced Study at Princeton, New Jersey, in the United States. Convinced of the inevitability of yet another war in Europe, Einstein abandoned his pacifist ideals and urged free Europe to arm for defence. On 2 August 1939, with a war in Europe imminent, Einstein (persuaded by Leo Szilard, Eugene Wigner and Edward Teller, three physicists who had fled Europe to escape fascist persecution) wrote the fateful letter to President Franklin D. Roosevelt which set in motion the age of nuclear arms. After the Second World War, Einstein championed the unrealistic notion of a world government to control nuclear weapons. This brought him a great deal of public acclaim but total indifference from those who really mattered, politicians and statesmen.

In 1950 Einstein presented a revised version of his unified field theory. This was a mathematically meticulous analysis but most physicists rejected it as untenable. Einstein died in his sleep (on 18 April 1955), after a lifetime in which he had attempted to influence both the affairs of humanity and of science. His crusade in human affairs had no lasting impact but his influence on science was profound. He changed, forever, human perception of the universe.

Although the legend of Einstein and his theory was growing in the first quarter of the twentieth century, the actual science of general relativity was becoming stagnant and sterile. Karl Schwarzschild (1916), Ludwig Flamm (1916), Hans Reissner (1916) and Gunnar Nordström (1918) discovered elegant solutions of Einstein's field equations within a few months of Einstein's presentation of general relativity. Hidden in these solutions were such bizarre entities as neutron stars, X-ray binaries, black holes[7] and wormholes, but these were not recognised at the time. More outlandish schemes to unify Einstein's theory of gravity and Maxwell's theory were proposed by Theodor Franz Kaluza (1919) and Oskar Klein (1926) in a five-dimensional theory of gravity. Taking their cue from Einstein, the two mathematicians, working independently, sought to show that the electromagnetic force might be accounted for by a fifth dimension. To explain why the effects of the fifth dimension were not visible at normal energies and distances, they assumed that the fifth dimension was rolled-up so tightly that it was smaller than an atom. In the Kaluza–Klein theory, each point of normal space is a loop in this fifth dimension. A charged particle at rest in normal space is in continual motion round this loop. In this theory the electric charge is actually motion in this hidden dimension. There are a number of connections between this motion and classical electromagnetism; for example by applying Newton's Third Law of motion – for every action along the rolled-up dimension there is an equal and opposite reaction – it is possible to get the law of conservation of electric charge. Remarkably Kaluza and Klein had succeeded in unifying Maxwell's electromagnetic theory with Einstein's theory of gravity. However, further detailed study of this theory revealed a number of inconsistencies and most physicists viewed the theory with scepticism. The full power of the Kaluza–Klein approach was not appreciated until the 1980s. By the late 1920s Einstein had turned his considerable genius to the futile quest to enhance the geometric structure of space-time to encompass all known physical forces – a unified field theory. From time to time he also attempted to make sense of Kaluza–Klein five-dimensional theory. We now know that a synthesis of gravitation and electromagnetism was a wrong step towards a unified field theory. In 1967–68, Stephen Weinberg, Abdus Salam and Sheldon Glashow realised that the correct unification scheme was to unite electromagnetism with the theory of weak interaction, in what is now called the electro-weak theory (see Chapter 8, Planck).

After the 1920s general relativity was not a very active area of scientific interest. The impression was that general relativity was a fundamental theory of nature and of profound importance but it was difficult to understand, impossible to work with and had few observational consequences. But this view was to change in the later half of the twentieth century. The 1950s and 1960s were times of change, when old and accepted conventions were once again being challenged. Young theoreticians were willing to take up the challenges posed by general relativity. These young minds were sufficiently flexible not to be fazed by the bizarre solutions of Einstein's field equations. In addition, the scientists released from defence establishments, both in the USA and in the former USSR, were able and willing to apply the tools they had developed for weapons design to general relativity. New astronomical discoveries being made at the time, like the expansion of the universe, neutron stars, black holes and quasars could only be understood within the context of general relativity. This was also the time when viable and attractive alternatives to general relativity were being proposed (see *Alternatives to general relativity* later in this chapter). This spurred the design of new experimental tests to confront the alternative theories and challenge general relativity. Some of these experiments and their results are described in the next chapter.

CHOICE BETWEEN NEWTON AND EINSTEIN

Early in 1916 the German astrophysicist Schwarzschild obtained rigorous and exact mathematical solutions of Einstein's equation of general relativity for two special and simple cases. Einstein presented these solutions, on his behalf, to the Prussian Academy of Science in Berlin on 13 January 1916. Karl Schwarzschild (1873–1916) was an exceptional astronomer whose practical and theoretical contributions were of primary importance in the development of twentieth-century astronomy. He demonstrated his ability in science at the age of 16 with a paper on the theory of celestial orbits. By 1910 he was director of the Potsdam Astrophysical Observatory. He made major contributions in the classification of stars and was the first physicist to recognise the role of radiative processes in the transport of heat in the stellar atmosphere. He was also one of the pioneers in developing the atomic theory of spectra proposed by the Danish physicist Niels Bohr. While serving in the Imperial German army during the First World War he contracted a fatal illness and died on 11 May 1916.

Schwarzschild's solution of Einstein's general relativity equation was for a single isolated spherical body. He showed that for a body of given mass, an observer at a fixed distance from the centre of the body would measure the same curvature of space-time (due to the gravitational strength of the body) irrespective of the radius of the body. As the radius of the body decreased (or mass increased) the strength of the gravitational field around the body would increase and the space-time would become increasingly curved. At a certain critical radius of the body the gravitational field at the surface of the body could be strong enough to trap even light emitted from the surface of the body, and the body would be cut off from the rest of space-time. The radius of the body at which it becomes invisible, to a distant observer, is called the Schwarzschild radius (discussed later in this section). This is the foundation of the theory of black holes, which will be considered later. But if the mass of the body was 'small', so that the gravitational field at the surface was weak then the geodesic motion of a body in this field reduced to that described by Newtonian mechanics. This is not so surprising as Newtonian mechanics is very successful in describing motion in a weak gravitational field. The observations of the orbital motion of double stars, the dynamic motion of clusters of stars collectively moving in a galaxy and the motion of galaxies verify that Newton's law of gravity is valid to a high degree of accuracy throughout the visible universe. The trajectories of space probes to planets or to comets and the path of the space shuttle can be calculated with great precision with Newtonian mechanics. It is not surprising that one of the Apollo astronauts commented that 'Isaac Newton is doing most of the driving right now.' Newton was also doing all the driving when Halley's Comet reappeared, on schedule, in 1986. This was the scientific media event of the year and a number of space probes were launched a few months earlier to rendezvous with the comet, again driven by Newton.

Schwarzschild's solution demonstrated that in all situations of weak gravity and where the gravitating bodies move slowly compared with the speed of light, the theories of Newton and Einstein (almost) agree. To illustrate this it is worth considering the minimum speed a body of unit mass has to acquire to escape the gravitational attraction of a massive body; this is called the *escape speed* of the body (escape speed, in kilometres per second, of a few typical astronomical bodies is given in column two of Table 5.1). On the surface of the Earth the escape speed is 11.2 kilometres per second (about 40 thousand kilometres per hour). A gravitational field is considered strong if the escape speed approaches the speed of light. Thus if the ratio of escape speed to the speed of light is between 0.1 and 1.0 the relativistic effects are important. The value of this ratio for some astronomical bodies is given in Table 5.1.

This ratio can increase if the mass of the gravitating body increases and/or its radius decreases. When a body of fixed mass decreases in size, the relativistic effects become important when the radius approaches the *Schwarzschild radius*. The Schwarzschild radius of the Sun is about 3 kilometres. Stars can collapse to a radius of this order in the late stage of evolution – such stars are called *neutron stars*. The density of

Table 5.1. *Escape speeds for a few representative astronomical bodies*

	Escape speed (km s^{-1})	Ratio of escape speed to speed of light
Moon	2.4	8.0×10^{-6}
Earth	11.2	3.7×10^{-5}
Jupiter	60.8	2.0×10^{-4}
Sun	640	2.1×10^{-3}
Sirius B	4800	0.016
Neutron star*	115 181	0.38

* A neutron star of 1.0 solar mass and radius of 20 kilometres has been assumed.

matter in a neutron star is enormous: for a one-solar-mass neutron star the density is of the order of 10^{17} kilograms per cubic metre. This density is comparable to the density of nuclear matter. If a star collapses to its Schwarzschild radius and continues to collapse further (if its mass is sufficiently large) then it becomes a *black hole*. Neutron stars and black holes are described later in Chapter 7 (Hubble & Eddington).

The critical density at which a body reaches the gravitational radius depends only on the mass of the body. A cluster of 10^8 stars can collapse to a black hole if the average density of matter in the cluster approaches or exceeds the density of water (1 gram per cubic centimetre). This seems to happen in the nuclei of some galaxies, and these *active galaxies* or *quasars* will also be described in Chapter 7. Relativistic effects can also become important if the size of a body increases but the density stays fixed (or the total mass increases). For such bodies the relativistic effects are important if the ratio of mass to radius is greater than about 1.3×10^{27} kilograms per metre. This limit is impossible for most bodies of realistic size and mass. There is, however, one body whose size can be as large as we like, namely the universe. The mean density of the observable matter in the universe is about 10^{-28} kg m^{-3}. A sphere of radius of the order of 10^{27} m would enclose enough matter to reach the critical limit and become relativistic. Such sizes are in the realm of cosmology – the study of the universe as a whole. One of the major achievements of the theory of general relativity was its ability to describe the universe in a consistent manner. This was not possible in the context of Newtonian physics as shown in Chapter 4, Newton. The formulation of the general properties of the universe in the context of the theory of general relativity will be considered in Chapter 7, Hubble & Eddington.

ALTERNATIVES TO GENERAL RELATIVITY[8]

The search for a field theory of gravitation had begun with Maxwell. In his great memoir *A Dynamical Theory of Electromagnetic Field*, completed in 1864, Maxwell noted the similarity of the inverse square law of gravitational attraction to electric and magnetic attraction and repulsion. This led Maxwell to wonder if gravitational attraction was 'not also traceable to the action of the surrounding medium'. However, gravitational force is *only* attractive whereas the electric force can be both attractive and repulsive. This implies that the presence of a dense body diminishes the energy of the surrounding medium. Maxwell was unable to see how a medium could possess such a property and he abandoned the search for a field theory of gravitation.

The basis of the relationship between space, matter and gravitation, which was linked into the theory of general relativity by Einstein, may have been anticipated 45 years earlier by the British mathematician William Kingdom Clifford (1846–1879). Clifford entered Trinity College, Cambridge, in 1864 at the age of 18. He excelled in mathematics, philosophy and literature. Eight years later he was appointed to the chair of mathematics and mechanics at University College London. In 1873 he translated, for the scientific journal *Nature*, Riemann's famous lecture *On the hypotheses which lie at the foundation of geometry*. He developed a number of Riemann's ideas and he speculated that the curvature of higher-dimensional space might appear like the force exerted by electricity and magnetism. Clifford extended these ideas to gravitation; he may have been encouraged by his colleague

Maxwell to pursue this line of inquiry. In a paper[9] published in 1876 he states

> the variation of the curvature of space is what really happens in that phenomenon which we call the motion of matter.

Unfortunately Clifford was unable to develop his theories further. On 3 March 1879, aged just 33, he died (like Riemann) of tuberculosis.

Between 1912 and 1913 the Finnish physicist Gunnar Nordström formulated the first logically consistent relativistic field theory of gravity. Nordström treated gravity like electromagnetism, as being due to a force field within Minkowski's flat, special relativistic space-time. Although the theory did not survive it had a number of attractive features. For example, in this theory the velocity of light was a universal constant as required by the theory of special relativity. This constancy of the velocity of light was violated by most other attempts at the time to formulate a relativistic theory of gravitation. Also the theory was Lorentz invariant and satisfied the conservation laws. But the theory did not rigorously satisfy the equivalence principle.

In the early 1960s the British astrophysicist Fred Hoyle and the Indian cosmologist Jayant Narlikar, working at the Institute of Theoretical Astronomy, Cambridge, proposed a theory of gravitation which was equivalent to Einstein's general relativity in the description of all observable phenomena, but appeared to have a number of satisfying features. For example, it required a positive gravitational constant G, whereas a positive G is introduced empirically in both Newton's and Einstein's theory of gravitation. Also the magnitude of G followed from the mean density of matter in the universe. Stephen Hawking, just starting on his brilliant career as a theoretical astrophysicist, found the theory attractive but inapplicable to the observed universe. Moreover, further critical analysis seemed to suggest that the theory did not unambiguously lead to a positive G and failed the classical tests of general relativity.

Any theory of gravity has to agree with two fundamental bodies of experimental knowledge. Firstly, the new theory's nongravitational formulation (that is the formulation that results when gravity is 'turned off' or when the effect of gravity can be somehow ignored) must agree with Einstein's special relativity. The validity of special relativity in the 'absence' of gravity has now been proven to a very high degree of precision in high-energy particle accelerators. Secondly, the new theory has to agree with Newtonian theory: that is, its predictions should be compatible with the observed properties of planetary motion. These two criteria have ruled out a number of theories of gravity proposed in the 1950s and 1960s. Almost all of the remaining theories can be divided into two types: metric theories and nonmetric theories. In a metric theory gravitation is a consequence of the curvature of space-time. In these theories events being observed take place in non-Euclidean four-dimensional space-time. Einstein's theory of general relativity is a metric theory. Different metric theories differ in the way in which matter (or energy) causes the space-time to be curved. Nonmetric theories, on the other hand, rely on other mechanisms to produce gravity: for example the quantum theory of gravity, if and when it is formulated, will (most probably) describe gravitational attraction in terms of exchange of an elementary particle called the graviton.

To distinguish between metric and nonmetric theories it is necessary to turn to experiments. The most powerful test of the viability of a metric or a nonmetric theory is: can it satisfy the equivalence principle? Experiments to test the equivalence principle will be described in detail in the next chapter. In the late nineteenth century the equivalence principle was tested to high precision by the Eötvös experiment. For theories of gravitation the role of the Eötvös experiment is similar to that of the Michelson–Morley experiment for special relativity. This experiment has now been performed to a very high precision and the results, combined with the conjecture of Leonard Schiff, that a viable theory of gravity that agrees with the equivalence principle is a metric theory, rule out a number of nonmetric theories. Higher precision experiments to test the equivalence principle are planned (see Chapter 6, Dicke) and these will constrain the nonmetric theories that have survived or that may be proposed in future (e.g. theory of quantum gravity).

One of the strongest alternatives to general relativity is the Brans–Dicke theory. Robert H. Dicke, together with his student Carl H. Brans, developed this theory in 1961. In developing this theory Dicke was influenced by Mach's principle (see Chapter 6). The theory predicted, in contrast to general relativity, that the gravitational constant G should vary with time. The theory also made a number of geophysical and astronomical predictions, which were discussed extensively in the 1960s. These predictions depend on a parameter which was introduced in the theory in an *ad hoc* manner. The value of this parameter can range from $-3/2$ to infinity. At large values of this parameter the theory resembles the general theory of relativity. In many ways this theory triggered a renewed interest in general relativity and theories of gravitation generally. The choice between various metric theories, including the theory of general relativity, is only possible by comparing the predictions of these theories with experimental results. In the solar system experiments (see Chapter 6, Dicke) the metric theories predict similar observable effects but the size of the predicted effect depends on the parameters of the theory. It is therefore necessary to continue improving the precision of these experiments because their results pin down the true value of the parameters. It is also necessary to devise new experiments, both terrestrial and astronomical, to test theories of gravity. It is interesting to note that up to now the results of all experiments (which unfortunately are not many) to test theories of gravity have been found to be in complete agreement with Einstein's theory.

CONSEQUENCES OF GENERAL RELATIVITY

Our deepening understanding of the theory of general relativity since its formulation has led to the identification of a number of bizarre and esoteric consequences of this theory. Some of these have since been identified with real physical objects or mechanisms but some still remain in the realm of speculation. These are briefly described here, and some will be considered in greater detail in the following two chapters.

Frame dragging or gravitomagnetic effect
Just as a moving electric charge will create an electromagnetic field, a moving body will create a gravitational field in addition to the field created by its mass. This additional field is due to the momentum of the moving body. This field can couple to the motion of other bodies (just as the electromagnetic field of a moving charge can couple to other electric charges) – this is called a *gravitomagnetic effect*. An important example is rotation. A rotating body like the Earth has angular momentum, which generates a 'gravitational field' that is proportional to the angular momentum. Because of this additional 'gravitational field' a particle falling freely in the Earth's gravitational field will acquire a small motion in the direction of the Earth's rotation. Also, a clock slowly co-rotating around a spinning body will advance relative to a clock which is at 'rest' (with respect to distant stars). This is called *'frame dragging'* or the *Lense–Thirring effect* after its discoverers. This dragging can affect the orbital period of satellites; the period of a satellite orbiting from west to east will be slightly less than that of one orbiting in the opposite direction (at the same altitude). Similarly a spinning gyroscope in orbit around the Earth will be 'dragged', resulting in the precession of its axis (a gyroscope is a rotating wheel or sphere whose axis is free to turn but maintains a fixed direction, relative to distant stars, unless perturbed – for more about gyroscopes, see Chapter 6, Dicke). This small and technologically challenging effect provides an opportunity to test the validity of general relativity and is part of an active programme of research.

Black holes and singularities
The Schwarzschild solutions of Einstein's field equation established that the 'gravitational stress' in nonrelativistic bodies like the Sun is small but in relativistic (that is, high mass and small radius) bodies the

consequences can be catastrophic. The dominant stress inside a star is the ordinary gas pressure. In a relativistic object like a neutron star, the gas pressure and the density of the stellar material are comparable. Such a large density increases the effective gravity inside the star. In general relativity there is a limit to how compact a star can be; no star of mass M and radius R can have $GM/Rc^2 > 2.25$. Suppose a star is at this very limit and (somehow) its radius is decreased slightly. The star will not be able to restore the equilibrium (it cannot stabilise at a smaller radius because this will violate the above inequality) and it will collapse. This is called a *gravitational collapse*, and the star will become a *black hole*. This is a consequence of the positive curvature of space-time implied by the attractive nature of gravity. What it means is that matter can curve a region of space-time on itself so much that it can effectively cut itself off from the rest of the universe. This was established in 1931 by the Indian-American astrophysicist Subrahmanyan Chandrasekhar and in 1939 by the American physicist J. Robert Oppenheimer. Black holes are described in greater detail in Chapter 7 (Hubble & Eddington).

The mass that has collapsed to form the black hole cannot find an equilibrium radius and will continue to move inwards. The British mathematical physicists Roger Penrose and Stephen Hawking have shown that this collapsing matter will always generate some form of *singularity* – a region of space-time where the matter that once composed the star is crushed out of existence (matter-free mass remains). At singularities the general theory of relativity breaks down. Singularities are an inevitable consequence of general relativity and they are a challenge to Einstein's theory. This is because space-time comes to an end or begins at these singularities. The nature of these singularities is poorly understood and this is an active subject of research. Some physicists comfort themselves with Penrose's 'cosmic censorship' conjecture that singularities in black holes are confined and (maybe) harmless, and naked singularities are forbidden. But this conjecture remains unproven. Others believe that the laws of quantum gravity, also ill understood at present, will deepen our understanding of these singularities.

Time machines and wormholes

In 1949, the famous logician Kurt Gödel found a model of the universe that satisfied Einstein's equations of general relativity and in which time travel was possible. Gödel was a strange man. He was born in 1906 a year after Einstein published his theory of special relativity, and was a close colleague of Einstein at the Institute of Advanced Studies in Princeton. A hypochondriac and a depressive, he developed a profound distrust of doctors and food, which he believed, was poisoned. He died of malnutrition in December 1977. Gödel's universe neither expanded nor contracted but it rotated. In this universe it was possible to travel backwards in time just by going out to a great distance from Earth and then returning. Time travel has bizarre and lurid implications, for example autoinfanticide – where one (accidentally) kills oneself as an infant. Gödel's universe and the associated time travel were not taken seriously, as the observed universe is very different from that described by Gödel's solution. The observed universe does not appear to spin and it is expanding. Even before Gödel, J. van Stockum, working in Edinburgh in 1937, had discovered a solution to Einstein's equations in which an infinitely long, rapidly spinning cylinder functions as a time machine. But infinite cylinders are not practicable and van Stockum's time machine has no practical value.

Time travel regained partial respectability in 1988 when three American physicists, Michael Morris, Kip Thorne and Ulvi Yurtsever, showed that by taking very generous (and perhaps unrealistic) liberties with the quantum properties of matter, stable *wormholes* might be possible. A wormhole is a hypothetical tunnel joining two points in a curved space-time; this is illustrated crudely in Figure 5.9. If our universe is curved (we would not be aware of it, just as an ant crawling over a sheet of paper is unaware of the gentle folds in it) and if stable wormholes exist or can be produced then they can provide a shortcut between different parts of the universe. Wormholes 'enable' travel at speeds faster than light and also travel backwards in time. Ludwig Flamm had recognised wormholes in 1916 in Schwarzschild's solutions of Einstein's equation. They were studied extensively in the 1950s by the

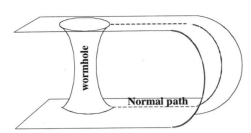

FIGURE 5.9 A schematic diagram of a wormhole connecting two distant regions of a universe, if the universe has a fold in it. The wormhole provides a path between two locations which is shorter than the normal path. This is an embedding diagram which idealises the universe to a two-dimensional sheet.

American physicist John Wheeler and his research group. These early wormholes were unstable, lasting for such a short time that no signal, let alone a person, could travel through them. Kip Thorne became interested in wormholes in 1984–85 (in his book *Black Holes and Time Warps* Thorne states that he developed an interest in wormholes when he was looking for a scientifically respectable basis for the science-fiction novel *Contact*, written by his friend Carl Sagan), and this interest led to the 1988 paper demonstrating that stable wormholes were possible. There is, however, considerable uncertainty about wormhole physics at present and quantum gravity effects may well prevent a wormhole being converted into a time machine. The British mathematician Stephen Hawking has proposed a *chronology protection* conjecture, which states that *laws of physics do not allow time machines* (or as Hawking would have it, 'keep the world safe for historians'[10]).

Gravitational lenses

The deflection of a light in gravitational field was confirmed in 1919 by the eclipse observations of Dyson and Eddington (see Chapter 6, Dicke). It was realised soon afterwards that the light from a source has different paths in the non-Euclidean space-time close to the deflecting mass (Figure 5.10) and this could result in multiple images of the source, or gravitational lensing. Lensing of galactic stars was expected but none was observed and interest in gravitational lenses waned. In 1960s Jeno Barnothy and S. Refsdal independently suggested that

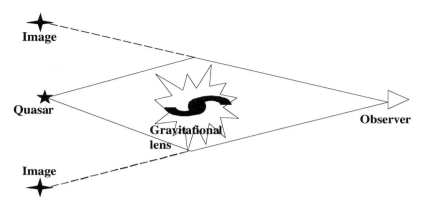

FIGURE 5.10 Gravitational lens. Light from the quasar is bent round the lens (a galaxy or a cluster of galaxies) to produce multiple images at the observer. The light beam bends because of the large curvature of space-time around the lens.

extragalactic objects might be lensed and these might be observable. But their suggestions did not cause a great deal of excitement in the astronomical community and the gravitational lenses remained amusing curiosities for the next few years. The situation changed dramatically on 29 March 1979 when a team of British and American astronomers (Dennis Walsh, Robert Carswell and Ray Weymann) obtained the optical spectra (at the Kitt Peak National Observatory, USA) of two radio-loud quasars about 6 arcseconds apart on the sky. Quasars themselves are a consequence of the strong gravitational field and will be described in detail in Chapter 7 (Hubble & Eddington). What was unusual about the two quasars, designated 0957 + 561 A and B, was that their optical spectra were almost identical. Both the number and redshifts of the emission and absorption lines in the spectra of two quasars were similar. The similarity of the two quasars suggested that either Walsh and his colleagues had observed two very similar (twin) quasars, or they had observed *two images of a single quasar which was lensed by an intervening gravitating mass.* Walsh and his colleagues opted for the gravitational lens. In the following months the lensing galaxy and the associated cluster of galaxies were discovered when deep CCD images of the quasar field were obtained

(CCD stands for charge coupled device, a means of electronic imaging). Additional data at radio frequency and at ultraviolet wavelength were also obtained; the later with the newly launched ultraviolet space observatory, the International Ultraviolet Explorer (this satellite was a joint project undertaken by NASA, European Space Agency and the United Kingdom). These observations proved that the two images were achromatic, an unambiguous signature of a gravitational lens (because the lensing gravitational field will equally affect photons at all frequencies, and therefore the ratio of intensity of the lensed images will be the same – within observational limits – at all frequencies). These observations confirmed what Walsh and his colleagues had believed all along, that 0957 + 561 A and B were gravitationally lensed images of a quasar.

The discovery of Walsh and his colleagues has triggered considerable interest in gravitational lenses in the past 20 years and these lenses are proving powerful tools for cosmological studies. About 50 examples of strongly lensed systems are now known. An example of images of a quasar caused by the lensing effect of a massive foreground galaxy is shown in Figure 5.11. These systems enable astronomers to map the distribution of mass in the lensing object. They also allow astronomers to observe small and faint features in distant galaxies, because the features are magnified by the lens effect. Several star-forming galaxies have been observed in great detail with the help of these naturally occurring magnifying glasses. These multiple images can also be used to determine the global cosmological parameters of the universe such as the mass density, the cosmological constant and the expansion rate of the universe. The first two parameters determine the overall geometry of the universe, that is, how the real distance of an object is related to its measured redshift.

The expansion rate of the universe, called the Hubble constant H_0, is a crucial cosmological parameter. The conventional method of determining its value is to construct a 'cosmological ladder' based on the relation between period and luminosity of a class of stars called Cepheid variable stars (see Chapter 7, Hubble & Eddington). The gravitationally

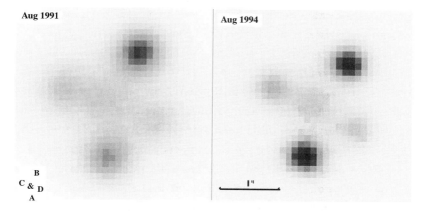

FIGURE 5.11 'Einstein Cross' – four images of a background quasar in line with a massive galaxy, whose image is in the centre. The gravitational field of the galaxy creates the multiple images. The differences in the brightness of the four lensed images in the two pictures (obtained 3 years apart) are caused by stars in the lensing galaxy acting as mini-lenses and magnifying the images individually. A William Herschel Telescope (La Palma Observatory) image obtained by G. Lewis (Institute of Astronomy) and M. Irwin (former Royal Greenwich Observatory). Material created with support to AURA/STScI from NASA contract NAS5-26555.

lensed images provide an alternative method of determining H_0. In a multiply lensed system if the 'object' quasar varies in brightness then the time taken by the change to travel along different paths will be different. There will thus be a delay in the variation of brightness of the images. The time delay depends on two factors, the difference in the geometrical path lengths and the difference in the gravitational potential along the paths. If the time delay can be measured and the gravitational potential of the lens is known or can be modelled then the Hubble constant can be determined. Unfortunately the task of modelling the gravitational potential of a lens is not trivial. The observable parameters of galaxies and the clusters of galaxies provide limited sets of constraints for the models, leaving astronomers to make assumptions about the shape and size of galaxies and clusters. Because both observational and theoretical aspects of the problem are challenging, progress in determining the Hubble constant in this way has been rather slow. In principle the method is very powerful as it enables the Hubble constant to be measured over truly cosmological scales.

Gravitational radiation

The instantaneous action at a distance of Newtonian gravity implies that the gravitational effects propagate at infinite speed. The theory of special relativity, on the other hand, imposes a strict upper limit on the speed of communication of physical effects; this limit is the speed of light. In a paper published in 1916, Einstein showed that the equations of general relativity permit gravitational radiation with the characteristic speed equal to the speed of light. Einstein also showed that these waves carry away energy from the radiating system. In the early days of general relativity, many physicists, including Einstein, believed that gravitational waves were a mathematical artefact, which would go away after more rigorous analysis. But work in the 1950s and 1960s established the reality of these waves and confirmed that they would transfer energy and angular momentum away from the source.

Gravitational waves are ripples in space-time that travel at the speed of light. The details of how gravitational effects propagate are very complicated, especially in situations where the gravitational field is strong. But for a weak gravitational field the gravitational waves are remarkably similar to electromagnetic waves. They differ in one fundamental aspect: unlike electromagnetic waves, gravitational waves can neither be scattered nor be absorbed. They can be gravitationally lensed but otherwise they propagate unhindered. They thus carry information about an event and this information is not corrupted along the path of propagation; the waves therefore have enormous diagnostic potential. Electromagnetic radiation is emitted by an oscillating electric charge (Figure 5.12(a)). This radiation moves away at the speed of light. The electric and magnetic disturbances carry energy generated by the oscillating charge; as this charge loses energy, its motion is damped and will eventually cease. The gravitational analogue of an oscillating charge is a dumbbell rotating along an axis perpendicular to its handle (Figure 5.12(b)). This system is similar to a binary star system

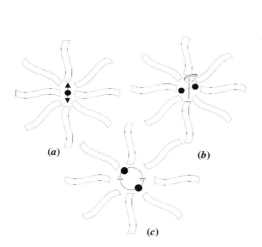

FIGURE 5.12 Generation of radiation waves. Electromagnetic waves are produced by an oscillating electric charge (a). The analogue for production of gravitational waves is two rotating masses (b). The astronomical equivalent is a binary star system (c). As the stars rotate the emitted gravitational waves carry away the energy of the binary system. Because of loss of energy the stars will spiral towards each other or the period of the binary system will increase.

(Figure 5.12(c)). The rotating dumbbell will generate disturbances in space-time, which will propagate at the speed of light. These disturbances will also carry energy, which will cause damping of the rotating system slowing down the rotating dumbbell, or will cause the binary stars to move closer and their angular speed will increase.

A dramatic confirmation of the reality of gravitational waves was provided in December 1978 (at the start of the centenary year of Einstein's birth) when Joseph Taylor presented the results of observations of the binary pulsar PSR 1913+16 at the Ninth Texas Symposium held in Munich, Germany. Pulsars were discovered in 1967 by two radio astronomers working in Cambridge, England (see Chapter 7, Hubble & Eddington). By 1974 over 100 pulsars had been discovered and sky-surveys were being conducted to create a larger database to build a statistical picture of pulsar properties. The American astronomers Taylor and his research student Russell Hulse, from the University of Massachusetts at Amherst, had an observing programme at the 1000-foot Arecibo Radio Telescope in Puerto Rico to survey the sky for more pulsars. On 2 July 1974, Hulse noticed an unusual weak object. The pulsar had a very short pulse period, only 0.059 seconds.

Only one pulsar was known which had a shorter pulse period: the Crab pulsar in the Crab nebula, which is a remnant of a star that exploded in 1054 AD and was extensively monitored by astronomers, mostly in the Far East. On 25 August Hulse decided to take a second look at this unusual object. In a 2-hour observing run he measured two *different* pulse periods of the pulsar. This was totally unexpected, as pulsar pulse periods were known to be very stable. Hulse concluded after another month of observations that the unusual object he was observing was a binary system with an orbital period of 7.7522 hours. Different pulse periods were observed because these observations were made when the pulsar was in different parts of its orbit and was moving at different speeds, and the frequency of the pulses was Doppler-shifted. Hulse and Taylor determined that the orbit was highly elliptical and tilted. More interestingly, they determined that the pulsar was moving at about 300 kilometres per second (about one-thousandth of the speed of light) and the circumference of the orbit was almost equal to the circumference of the Sun. This was a 'relativistic situation' and sent a buzz through the community of relativists when it was reported in the *Astrophysical Journal Letters* in January 1975.

As the observations of the pulsar continued in 1974, relativistic effects began to be apparent. In December 1974 Taylor reported that the periastron advance for the binary pulsar was 4 degrees per year – compare this with the 43 arcsecond per year perihelion advance of Mercury. Taylor and his colleagues then 'used' general relativity and very accurate timing techniques to determine that the masses of the neutron stars in the binary system were respectively 1.4411 and 1.3873 solar masses. These masses were in very good agreement with the theoretical estimates of the mass of a neutron star (see Chapter 7, Hubble & Eddington). But the most startling news was presented at the Texas Symposium in Munich – Taylor reported that the orbital period of the binary system was decreasing; the two neutron stars were spiralling towards each other. This was proof that the binary was losing energy. By 1990 Taylor and his colleagues had determined that the decrease in the orbital period was 76 ± 2 milliseconds per year. This is in excellent agreement with the 75 milliseconds per year decrease predicted by the theory of general relativity. At present, this is the only observational evidence for gravitational waves. Although the evidence is circumstantial, the agreement with theory is so good that there is really no doubt about the reality of gravitational waves. In 1993 Hulse and Taylor were awarded the Nobel Prize in physics for their discovery of the binary pulsar and the determination of relativistic effects in this pulsar. In 1991 a second short-period binary pulsar was discovered, PSR 1534 + 12, also by Taylor's group. Both binary pulsars have been timed very accurately over a long enough period to provide significant tests of general relativity in a strong field limit (unlike the solar system tests, which are in a weak field). These observations have provided experimental proof that strong field binary systems lose energy by emitting gravitational waves and these waves propagate at the speed of light.

6 Dicke

Einstein was motivated by a deep philosophical need, the quest for simplicity and unity in nature, to formulate and develop the theory of general relativity. He was not guided by a desire to confirm or interpret any particular experimental result(s) although he was aware of the need for experimental confirmation. Experiments are fundamental to modern physics: progress in physics is driven by experimental verification and no assumption can be taken seriously unless it can be tested experimentally. This is the only way to distinguish physics from metaphysics. Galileo repeatedly stressed this and his experiments in the sixteenth century were able to overthrow the 2000-year reign of the speculative laws of nature proposed by Aristotle. Today a theory without experimental verification has no value. Unfortunately general relativity, unlike its contemporary, quantum theory, does not have a secure experimental foundation. Einstein had shown that the perihelion shift of Mercury could be explained by general relativity with remarkable accuracy. He also proposed the gravitational redshift and the bending of light rays as two further tests of general relativity. Gravitational redshift was too small to be observed with the technology of the first half of the twentieth century. Also, as will be discussed later, this is really a test of the equivalence principle and not of the full theory of general relativity. The bending of light was measured in 1919 but the accuracy of the data was low and not sufficient to discriminate between general relativity and the alternative theories of gravity proposed in the 1960s. Similarly, there was considerable uncertainty, until recently, about the oblateness of the Sun which affects the perihelion shift of Mercury. Remarkably only one new test of general relativity has been proposed since the formulation of the theory by Einstein.

High precision experiments are essential to test and confirm general relativity and also to distinguish it from the competing theories that have been proposed since the 1960s or may be proposed in the future. This has forced unprecedented improvements in instrumentation and measurement techniques. Advances in technology have now provided the tools to test Einstein's theory to a precision that would have been unimaginable just a few years ago. The advent of the space age has opened up new avenues to test theories of gravitation in environments that are impossible in ground-based laboratories. These challenges of high precision and space-platforms have brought together disciplines that only a few years ago would have been considered totally incompatible. Gone are the days when one or two physicists got together and proved a fundamental fact or made groundbreaking discoveries. Today's experimental groups have physicists, mathematicians, mechanical engineers, electronic experts, computer whiz-kids and every conceivable technical expert, plus managers and accountants! This convergence of disciplines has benefited everyone involved; science has gained from the expertise developed and available in other fields. Also, 'nonscientists', who in the past would have only been interested onlookers, can now share in the thrill and excitement of fundamental discoveries.

In this chapter the experiments which have been performed to test the general theory of relativity and to distinguish between different theories of gravity are described. In every case the results have been found to be in complete agreement with Einstein's theory. Thus measurements of even higher precision are required to challenge the general theory. A brief description is also given of instruments and experiments which have been proposed to further and deepen our understanding of Einstein's relativity and gravitation generally.

UNIVERSAL CONSTANT OF GRAVITATION G

The essence of Newton's universal theory of gravitation is that the force between two bodies is proportional to the product of their inertial masses and the inverse square of their separation. The force depends on nothing else. In a weak gravitational field the same is also true of general relativity, with a small geometric modification. During the nineteenth century a number of experiments were performed to show that the gravitational force between two bodies was independent of temperature, electromagnetic field, shielding by other bodies, orientation of crystal axes, and other factors. In the Newtonian theory the constant of proportionality between the gravitational force and the mass and separation of the gravitating bodies is the universal constant of gravitation, G. In Einstein's theory of general relativity the constant G determines the amount of curvature of space-time produced by a given mass (or energy). There has been a continuing interest in determining the absolute value of the constant G, the constancy of this constant and the confirmation of the inverse square law. In many ways the constant G is rather anomalous compared with other constants of physics. It does not depend on the material properties of a body and should really be considered a geometric factor. Also, at present, G cannot be related to any other physical quantity and it is not possible to predict its value from other constants of physics (but see Chapter 8, Planck). Gravitational force is also different from other forces of nature; for example, electrostatic force can either attract or repel, depending on the charge of the bodies, but gravitational force can only attract, there is no gravitational repulsion. It is thus impossible to isolate a body gravitationally, as it cannot be shielded from the gravitational influence of surrounding bodies.

Gravitational experiments are notoriously difficult to perform. Gravitational forces have to be detected mechanically, and although the accuracy of these methods has increased in recent times it does not approach that which can be achieved in electrical experiments. Mechanical experiments cannot be gravitationally isolated and because the gravitational force is very small, these experiments are very sensitive to external disturbances.

The absolute value of the universal constant of gravitation can be measured in four different ways: first, by comparing the pull of a large mass with the pull of the Earth; second, by measuring the attraction of the Earth on a test mass; third, by measuring the force between two masses; and fourth, by measuring the speed of a body falling in a gravitational field. The first approach was suggested by Newton and was first attempted in 1774 by the British astronomer Nevil Maskelyne, on the mountain of Schiehallion in Scotland. The British physicist John Henry Poynting developed the laboratory balance method during the late 1800s. All recent measurements have attempted the laboratory measurement of the force between two masses, using a torsion balance. The torsion balance was invented by Reverend John Michell, but he died in 1793, before he could use it. Michell bequeathed his balance to the eccentric British scientist Henry Cavendish (1731–1810) who modified the balance to measure G. Cavendish was the first son of Sir Charles Cavendish, himself an experimenter of some note. Not much is known of Henry Cavendish's early education but he entered Peterhouse College, Cambridge, in 1749. Three years later he left Cambridge without taking a degree and took up residence in London. One of the wealthiest men in England, he lived in secluded isolation, devoting his life to science. In 1766 he submitted to the Royal Society his first paper, entitled *On Factitious Airs* (as opposed to the 'natural' air). He had found that a highly 'inflammable air' was produced by the action of acids on certain metals – Cavendish had discovered hydrogen. He also demonstrated that the only product of combustion of the 'inflammable air' (hydrogen) and 'dephlogisticated air' (oxygen) was water, and he gave an approximately correct proportion of the two in water. Until his death Cavendish continued, in splendid isolation, his scientific investigations in chemistry and electrical phenomena. Much later when Maxwell edited his unpublished papers he noted that Cavendish had anticipated a number of electrical phenomena independently discovered later by Michael Faraday and others. Cavendish had evidently carried out his investigations to satisfy his own curiosity and did not see a need to publish his findings. Cavendish died in 1810 leaving behind a considerable fortune and a large stack of scientific manuscripts attesting to his diverse interests and amazing experimental ability. In recognition of the latter Cambridge University named the Cavendish Laboratory after him; Maxwell was its first director.

In 1798, when he was 67 years old, Cavendish performed the experiment to measure G. It is not known why Cavendish turned to this problem except that he had discussed it with Reverend John Michell. He modified the torsion balance by attaching two small lead spheres, 2 inches (about 5 centimetres) in diameter, to the ends of the horizontal arm. The arm was 6 feet (about 1.83 metres) long and was suspended by a thin wire, 40 inches (about 1 metre) long, inside a wooden box (Figure 6.1). Outside the box were two lead spheres, 8 inches (about 20 centimetres) in diameter, arranged as close to the small spheres as possible so that they could attract the small masses. The large spheres were alternately placed on either side of the small masses to deflect the small masses in opposite directions (see the plan view in Figure 6.1). The whole assembly was enclosed in a room to maintain a uniform temperature. Cavendish measured the deflection of the arm from outside the room, with a small telescope. From this deflection he obtained a value of G equal to 6.754×10^{-11} N m^2 kg^{-2} (Newton square metre per square kilogram). Almost all experiments done since to measure G have been refinements of this wonderful experiment and the value of G was not bettered for almost 150 years. Cavendish used this value of G to 'weigh the Earth' and obtain its mean density. His result showed that the Earth must have a central core much denser than the surface rock.

The Austrian physicist Carl Braun introduced a variation of the 'Cavendish method' to improve the accuracy of the measured value of G. He observed the change in the period of oscillation of the torsion balance when the attracting masses were placed close to it. Higher accuracy is possible because the period can be measured with far greater precision than can a small deflection. In 1982, two American physicists, Gabriel G. Luther and William R. Towler, used this method to make a significant improvement in the value of G. They obtained a value of 6.67259×10^{-11} N m^2 kg^{-2}.

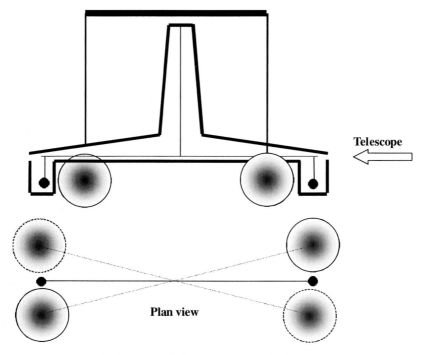

FIGURE 6.1 A schematic diagram of the torsion balance used by Cavendish to measure the gravitational constant G. The deflection of the suspended horizontal beam is measured when the two larger spheres are placed on either side of the two smaller spheres. The position of the large spheres is shown in the plan view. To maintain uniform temperature the whole assembly is kept in an enclosed room and the deflection of the beam is measured with a telescope.

At the Los Alamos National Laboratory in New Mexico, although officially 'retired' (he was born in 1932), Luther is continuing his attempts to improve the accuracy of G. To minimise external perturbations, the experiment is housed in a bunker in the desert about 20 kilometres from Los Alamos. His torsion balance is housed in a vacuum chamber and elaborate precautions are taken to keep the experimental assembly at a uniform temperature. Although the experimental arrangement for measuring G is simple compared with those of a number of other experiments in modern physics, the measurement itself is not trivial. A number of subtle effects have to be considered. Around 1996 Kazuaki Kuroda at the University of Tokyo pointed out that the classical torsion balance method assumes that the suspending fibre resists twisting with the same strength when the deflecting masses are near as when they are far. He claimed that this was not so and if this effect was not taken into account then the torsion balance experiment would over-estimate the value of G. Luther tested this and found that Kuroda was right: the G he had measured was slightly over-estimated. However, that was not the end of the story for the best value of G. The German standards laboratory in Braunschweig decided to avoid the Kuroda effect by floating their dumbbell on mercury. This allowed them to use a heavier dumbbell that would feel a stronger pull of gravity. They announced their results in 1994 and their value of G was significantly *higher* than the Luther–Towler value. More pain was to follow. The New Zealand Standards Laboratory also avoided the Kuroda effect by not allowing the pendulum to twist. They used an electrostatic method to stop the torsion pendulums from twisting when the deflecting masses were moved close to it. The electrostatic force required to prevent the twist can be measured very accurately and the value of G can be estimated from this resisting force. The value the New Zealand group determined was significantly *lower* than the Luther–Towler value. It is extraordinary that in an age when most constants of physics are known to an accuracy of six to eight decimal places, the value of the universal constant G is known at best to four decimal places. Not only is the value not known very accurately, but there may be unknown systematic effects in various experiments being made to measure it. These experiments are clearly going to continue. Future experiments are also likely to take to space to get away from seismic vibrations and other natural and man-made disturbances.

In celestial mechanics there is no independent method to determine the mass of a celestial body; the quantity that can be measured is the product GM. In laboratory measurements a mass is measured by comparing it to an arbitrary standard, the kilogram. The only way to relate the celestial mass to the laboratory standard is through the value of G.

The value of G is thus of fundamental importance to astronomy. The British physicist Paul A. M. Dirac (1902–1984) and others have suggested that G might be proportional to the age of the universe. Dirac was one of the pioneers of the quantum theory and he speculated widely on a number of other subjects including relativity and cosmology. His suggestion (speculation would be more accurate) was not based on any firm theory but in his belief that the physical world has a purpose. To Dirac the structures we see around us did not occur by chance. He arrived at this belief through what has come to be known as the 'large number hypothesis'. Briefly stated, the hypothesis is that it is possible to combine the absolute values of various constants of nature to produce numbers that are dimensionless (that is, in which the units cancel). Two such numbers are worth considering here: the ratio of the electric force between a proton and an electron to the gravitational attraction between the two, and the ratio of the age of the universe to the time taken by light to travel a distance equal to the classical radius of the electron (this is not equal to the true radius of the electron). Consider the ratio of forces first. Both the electrical and gravitational forces vary as the inverse square of the separation of the charged or gravitating particles, so the separation cancels in the ratio. The ratio thus depends on G, the mass of the electron and proton and the electric charge of the two particles. This ratio is of the order of 10^{40}. This value demonstrates that the electromagnetic force is considerably stronger than the gravitational force. Next, consider the age of the universe, estimated to be about 15 billion years or about 5×10^{17} seconds. This age is still not known very accurately (see Chapter 7, Hubble & Eddington) but the precise value is not relevant. The time taken by light to travel across the classical radius of the electron is about 10^{-23} seconds. The ratio of the age and the time of travel is also about 10^{40}. Dirac reasoned that the similarity of these two ratios was no mere coincidence, but that it pointed to a deeper and as yet unknown law of nature, which maintains this equality at all times. However, the age of the universe is not constant, it is continuously increasing, and therefore to maintain the equality of the two ratios one of the numbers involved in the ratios must change with time. There is a large body of experimental evidence that suggests that the electric charge, the mass of the electron and the proton, and the speed of light are real constants and independent of time. Thus the most likely parameter to vary with time appears to be G. A time-varying G (if it varies in the right way) would maintain the large value of the ratio of the two forces and the equality of the large numbers for all time. Dirac did not uncover the deeper law of nature whose existence he suspected, but the fascination with large numbers has not diminished.

Another reason for suspecting a time-varying G is Mach's principle. This principle is attributed to the nineteenth-century philosopher and physicist Ernst Mach (1838–1916). Mach was born in Moravia, part of the former Austrian Empire. His early education was at home. At 17 he entered the University of Vienna and was awarded a doctorate in physics in 1860. In 1864 he was appointed professor of mathematics at the University of Graz (Austria) where he became interested in psychology and physiology of sensation. Three years later Mach moved to the Charles University in Prague where he conducted experiments into the feeling associated with movement and acceleration. He also developed optical and photographic techniques for measurement of sound waves and wave propagation. In 1887 he established the principles of supersonic speed, and the *Mach Number* – ratio of velocity of an object to the velocity of sound – is named after him. In *Beiträge zur Analyse der Empfindungen* (Contributions to the Analysis of Sensation) published in 1887 Mach proposed that all knowledge is a conceptual organisation of the data of sensory experience or observations. This view leads to the proposition that no statement in natural sciences is admissible unless it is verified empirically. This rigorous criterion led Mach to reject such metaphysical concepts as absolute time and space, and prepared the way for Einstein's theory of relativity. Mach also speculated extensively on the nature of rotation, inertia and the 'fictitious' inertial forces arising in a rotating system. These have been central issues in mechanics since the time of Galileo. Mach believed that the centrifugal forces were the result of rotation relative to the mass in the universe. He also believed that inertial properties of objects do not originate from space but are determined (in an unknown way) by the existence of all other matter in the universe – Einstein called this 'Mach's principle'. Mach's principle is not a testable scientific principle (and therefore not an admissible proposition according to Mach's own rigorous criterion!) but a collection of Mach's thoughts on the nature of inertia and gravity.

Unknown to Mach and Einstein, the English philosopher George Berkeley had proposed similar views in the seventeenth century. Berkeley had argued that all motion, both uniform and nonuniform, was relative to the distant stars. Initially Einstein was fascinated by Mach's principle but in later years his enthusiasm declined because he realised that inertia is implicit in the geodesic equations of motion and need not depend on the existence of matter elsewhere in the universe. If Mach's principle is accepted then the gravitational force between two masses at a given separation must somehow be related to the distribution of matter in the universe. In an evolving universe the radius and the mean density of matter in the universe will change with time and therefore by Mach's principle the value of G must also change with time.

Both the large number hypothesis and Mach's principle suggest a time-varying G, in opposition to Newtonian theory and general relativity, which require a constant G. The time variability of G has recently been revived in the context of Kaluza–Klein and superstring theories (more about these in Chapter 8, Planck), which may unify the four forces of nature, namely the two nuclear forces, the electromagnetic force and gravitation. These theories naturally predict a variable G. Thus the determination of the variability of G may be a test of the existence of other dimensions and may make it possible to pin down the parameters of the superstring theory. The theories of varying G suggest that G should decrease with time or gravity should be getting weaker. If so, then stars, which are held together by gravity, should expand and the orbits of planets and satellites should also expand. This means that the length of the year or the lunar month should become longer. Because the possible variation is related to the evolution of the universe, it might be reasonable to assume that in the first approximation the value of G decreases at a rate corresponding to the rate of evolution of the universe. Since the universe ages at the rate of one year every year, it ages at the rate of a few parts in 10^{10} per year, if the age of the universe is about 10^{10} years. For this rate of decrease of G the length of the day would have increased by about 20% during the 4.5-billion-year age of the Earth.

One way to check for a variable G is to monitor the motion of the Moon and the rotation of the Earth. Observations of the motion of the Moon by monitoring the passage of the Moon against the background stars or by measuring the Earth–Moon distance by lunar laser ranging suggest that the lunar month is increasing at the rate of about 0.03 seconds per century or the lunar orbit is expanding at 2.6 centimetres per year. Similarly the length of the day, measured with atomic clocks, suggests that it is increasing at the rate of about 0.0016 seconds per century. Unfortunately these results cannot be interpreted unambiguously as decrease of G. The source of difficulty is the tides. Tides will be discussed more fully in the next chapter, but the net result is that tidal friction causes the Earth to slow down in its orbit and the Moon to recede from the Earth. Models of tidal friction suggest that the increase in the length of the lunar month or the length of the day is similar to the observed change within about 25%. Thus these observations do not tell us anything about the decrease of G.

The development of planetary radar ranging and atomic clocks has enabled very stringent limits to be put on the possible rate of change of G. The first such measurement was made by Irwin Shapiro and his group at the Massachusetts Institute of Technology (MIT), USA. Starting in 1964, this group timed the radar echo bounced off Venus (from 1964 to 1969) and Mercury (from 1966 to 1969) to measure the expansion of planetary orbits and showed that the rate of change of G was less than 4 parts in 10^{10} per year. The accuracy of this method was improved considerably by the radar ranging of Mariner 9 and the Viking spacecraft on and around Mars. This highly accurate measurement of the putative change in G was possible because of the high accuracy with which the distance to the lander or the Earth–Mars distance could be measured. An accuracy of 10 metres in the distance to Mars is possible. The unexpected 6-year life of Viking Lander 1 made repeated observations possible and allowed a high accuracy measurement of change in G possible. Staring in July 1976, 1136 ranging measurements were made between the Deep Space Network of the Jet Propulsion Laboratory, California, USA, and the Viking lander on Mars. The average interval between the measurements was about two weeks. The measurements were discontinued after July 1982 because by then the batteries on the lander were too weak for the transponders to function. The analysis of these data is horrendously difficult. Large amounts of data are involved which have to be corrected to a high degree of accuracy for perturbations of the orbits of Earth and Mars by other planets. The main factor limiting the accuracy of these results was the perturbing effect of the asteroids, the belt of interplanetary material that lies between the orbits of Mars and Jupiter. To calculate the net perturbing effect it is necessary to know the mass and orbit of every body in the belt, and these are simply not available (and may never be available). A lot of educated guesswork is involved in the calculation of the perturbing influence of the asteroids. Although the net effect is small it is nevertheless sufficiently significant to limit the accuracy with which the change in G can be measured. The result of two separate and independent analyses (one by the JPL group and the other by Shapiro's group at MIT) was that there was no evidence for a change in G to 1 part in 10^{11} per year.

More recently analyses of Lunar Laser Ranging (LLR) data obtained from 1969 to 1990 have resulted in accurate determination of a number of parameters relevant to the Earth–Moon system, the Sun and general relativity. The LLR can be said to be a 'near complete' gravity experiment because almost every parameter in a many-body relativistic equation of motion contributes to the measured perturbations of the Moon's orbit. 'Many-body' implies interactions of more than two bodies; for example the orbit of the Moon is *fully* determined not just by the gravitational field of Earth and the Moon but by the fields of the Sun, asteroids, etc. The LLR programme has its roots in the late 1950s and was the brainchild of Robert Henry Dicke (1916–1997). Dicke was born in St Louis, Montana (USA). He studied at Princeton University and the University of Rochester and in 1941 became a staff scientist at MIT. Five years later he moved to Princeton and stayed there for the rest of his working life. In 1975 Robert Dicke was appointed Albert Einstein professor of science. A larger-than-life personality, Dicke was not only at home with experimental and theoretical physics but he, exceptionally for the time, straddled the barrier between quantum theory and general relativity. In a career spanning almost 50 years he pioneered a number of ideas in radar technology, atomic physics, general relativity and cosmology. During the Second World War he worked at MIT developing the microwave radar and also developed the microwave radiometer, which has become an integral component of most modern radio telescopes. His work on microwave spectroscopy led him to formulate the first quantum theory of the emission of coherent radiation. This led later to the development of masers and lasers. In the late 1950s Dicke switched his attention to gravitation and cosmology. He carried out a number of high precision tests of the principle of equivalence (more about this later). In 1963, while investigating the cosmological consequences of his theory of gravitation (Brans–Dicke theory), Dicke postulated a relic background radiation of 40 degrees kelvin. But Dicke's universe was not the currently accepted big bang universe (described in Chapter 7, Hubble & Eddington), his was an oscillating universe. Dicke felt that an oscillating universe avoided the problem, in the big bang theory, of original creation of matter – or at least pushed it back to some remote past. Surprisingly Dicke (like others) was unaware that the existence of relic radiation of the primordial (big bang) fireball had been postulated 16 years earlier by three American physicists, George Gamow, Ralph Alpher and Robert Herman. Before Dicke could attempt to observe this radiation, Arno Penzias and Robert Wilson of Bell Telephone Laboratories discovered a faint glow of microwave radiation closely matching that predicted by theory, but the temperature of the background radiation was considerably lower than 40 K (see Chapter 7, Hubble & Eddington).

In the 1950s Dicke proposed measuring the change in the orbit of artificial earth satellites to determine the variation of G. This was to be achieved by measuring the change in the satellite's orbit by reflecting optical searchlight pulses off retroreflectors on the satellites. With the development of lasers in the 1960s it became possible to apply this technique to measure the distance to the Moon. With a laser it is possible to send very short and well defined pulses of light. The duration of the pulse determines the accuracy with which the round-trip travel time can be measured. During the first Apollo landing on the Moon on August 1969 a reflector was left on the lunar surface. Within weeks reflected pulses were detected at the Lick Observatory, California. The first measurement of the Earth–Moon distance was announced in 1976 and now it is possible to measure the distance with an uncertainty as small as 1 centimetre. These data established that the rate of change in G is less than one part in 10^{12} per year. This is about 1/35 of the Hubble expansion rate (see Chapter 7, Hubble & Eddington) of the universe. As more LLR data are accumulated the accuracy of these measurements will improve further. The LLR not only makes the measurement of variation of G possible, it also yields a greatly improved lunar orbit. This allows accurate study of the lunar libration, the angular oscillations about the centre of mass of the Moon. In addition LLR can give the precise separation between the light transmitting and receiving stations on the Earth. This can shed light on such questions as the wobble of Earth's axis, the movement of tectonic plates in Earth's crust and the lunar tidal acceleration. Some of the parameters of general relativity that can be measured from LLR data are described later in this chapter.

The solar system is a very good laboratory for high-precision tests of general relativity through observations of the Moon and the planets, and more importantly by tracking orbiting spacecraft and experimental packages landed on the Moon and the planets. More recently, precise millisecond binary pulsars have provided new laboratories to test general relativity. A binary pulsar consists of two neutron stars that are gravitationally bound and rotate around each other emitting gravitational radiation (recall the discovery of a millisecond binary pulsar by Taylor and Hulse, described in Chapter 5, Einstein). The observations of binary pulsars have a number of advantages for testing the variability of G (and for testing relativity generally): (a) the observations can be made over a very long period and are not limited by the life of the experiment or the mission; (b) a binary pulsar is a 'clean' astrophysical system with no perturbing matter surrounding it, unlike other binary systems; (c) a binary pulsar is a simple dynamical system of two gravitationally condensed bodies; and (d) the complete general relativistic theory of motion of such a system of two strongly self-gravitating bodies is well developed. The timing of the binary pulsar PSR1913+16 indicates that rate of change of G is less than 1 part in 10^{11} per year, which is consistent with the solar system observations.

Although the accuracy of these observations of the rate of change of G is very high it is only about a factor of 10 better than the naïve guess made above. It does not prove that G does not decrease with time. More accurate measurements are required. It is also worth bearing in mind that the rate of the putative change in G is measured relative to the time kept by atomic clocks. The change in G is believed to be due to cosmic effects coupled directly to gravitational physics. However, if the cosmic effects influence atomic physics, the atomic clocks may drift relative to the implicit clock of relativistic dynamics and this would appear like a change in G. If G were found to vary with time, at a rate comparable to the rate of expansion of the universe, the beautifully self-consistent fabric of astrophysics woven over the past 2500 years would begin to unravel. So watch this space – the full story of G has not been told yet.

INVERSE SQUARE LAW

The inverse square law is the second basic tenet of Newtonian gravitation and is also in agreement with Einstein's general relativity in the limits of weak field. On a laboratory scale, that is over a range of the order of a metre, the inverse square law was first tested and confirmed by Henry Cavendish. These experiments were done in the late 1700s with the torsion balance he had developed to measure the value of G. In the succeeding 200 years a number of refinements were made to this experiment and the inverse square law, on laboratory scales, has been confirmed to an accuracy of one part in 10^4. On the scale of a few hundred kilometres the inverse square law has been confirmed to one part in 10^6 from the data on planets and artificial satellites.

The interest in the inverse square law arises from the suggestion that the gravitational field itself may have a mass and the constants of gravitation may change over a characteristic scale related to the mass of this field. In 1986, interest in the inverse square law was further intensified when a fifth force (in addition to the four known forces of gravity, electromagnetism, weak interaction and strong interaction) was postulated by Ephraim Fischbach and his colleagues at the Purdue University and the Brookhaven National Laboratory in the USA. This group postulated a fifth force to explain the anomalous decay of some unstable elementary particles. The range of this suggested force was smaller than the range of gravity and would therefore manifest itself as a deviation from the gravitational attraction. An additional property of this force was that the acceleration caused by the force was expected to depend on the material of the body. This is a violation of the weak equivalence principle, which states that all bodies fall with the same acceleration. Recall that Newton had tested this experimentally. The weak equivalence principle has been tested to very high accuracy in experiments done between 1898 and 1908 by Baron Roland von Eötvös. In the 1950s, groups in Princeton and Moscow repeated these experiments with much higher precision. These experiments will be described later in this chapter. Eötvös and his collaborators measured the attraction, by Earth, of bodies of different material. They showed that the average attraction was equal to an accuracy of three parts in 10^9. Fischbach reanalysed the data of Eötvös and showed that the data actually indicated that the acceleration was different for bodies of different material. The acceleration appeared to be proportional to the ratio of baryon number and mass. Baryon number is the sum of the number of neutrons and protons in an atom. Fischbach and his colleagues concluded that this was evidence for a repulsive fifth force dependent on the baryon number. The strength of this force was comparable to the strength of gravity.

The acceleration by the fifth force may be distinguished from gravitational acceleration by: (1) violations of the gravitational inverse square law unless the scale length (the distance over which the force is effective) of the fifth force is infinite; and (2) violations of free fall. The two tests are partly complementary. The inverse square law tests are only weakly sensitive to forces with length scale greater than one astronomical unit. The free fall tests on the other hand are sensitive to a broad range of length scales. The inverse square law has now been tested over scale lengths from a few millimetres to 10^{12} metres. This has been achieved by experiments done in laboratories, in mineshafts, from tops of high towers, with artificial satellites in orbit around the Earth, and solar system tests done by measuring the orbits of planets by radar ranging and accurate measurements of the periods of planets. Over these scales the inverse square law has been found to be accurate to a few parts in 10^{10}. However, the law has not been established over galactic scales. Some astronomers have speculated that the 'missing mass' problem (this will be discussed in greater detail in Chapter 7, Hubble & Eddington) may indicate the breakdown of Newtonian gravity over cosmic scales rather than the presence of unknown exotic dark matter. But this speculation is inconsistent with the universality of free fall.

Experiments to test the universality of free fall can be traced back to Galileo and the so-called 'Leaning Tower of Pisa' experiments. These tests are done by simultaneously dropping two dissimilar objects. Recent interest in free fall was triggered by an Australian experiment in the early 1980s. The Australian group had measured the force of gravity down a kilometre-deep mineshaft in northern Queensland. The result was surprising: the net force measured at the surface was smaller than the force deep in the mine. Fischbach was aware of this experiment when he proposed the fifth force and he argued that the postulate of the fifth force provided a logical explanation for the observed discrepancy in the Australian experiment. Vertical free-fall experiments have set the most stringent limits on any violation of free fall for scale lengths between 20 kilometres and 1000 kilometres. The main difficulty in these experiments is that one is attempting to measure the difference in acceleration between two bodies moving at an acceleration of 9.80 metres per second per second. Potential sources of systematic error are innumerable. The results of the Australian experiment can probably be explained by gravitational gradients in the mineshaft in which the experiment was performed. Modern versions of Galileo's experiment have achieved a high degree of precision and the free fall of bodies of the same and different material has been tested to a precision of one part in 10^6. Torsion balance experiments, similar to that of von Eötvös, have shown that for a scale length of the order of one astronomical unit the acceleration of bodies of aluminium, gold and platinum is similar to a precision of 1 part in 10^{11-12}. But these experiments can miss interactions whose scale length is much smaller than one astronomical unit. An experiment at the University of Washington (USA) used a specially designed torsion balance to test the acceleration of different test bodies in the gravitational field of the Earth. A null result at the level of one part in 10^{11} was obtained for test masses of beryllium, aluminium and copper. More importantly these experiments did not find the correlation Fischbach had found between acceleration and the baryon number. The torsion balance experiments have also been performed close to navigational locks to test for effects of variable mass as a source of gravitational attraction. No effect was detected. More ingenious experiments with floating balls have given similar null results.

The present set of experiments reveal no composition-independent or composition-dependent deviation from the inverse square law over a number of length scales. This seems to rule out a fifth force of the sort postulated by Fischbach. But does this mean the inverse square law of gravitation has been established beyond doubt? In experimental physics 'beyond doubt' usually means beyond the imagination of physicists. This is not true for the inverse square law because nearly all attempts to extend the present framework of physics (supersymmetry, string theories – see Chapter 8, Planck) predict the existence of new interactions violating the universality of free fall. Thus higher precision tests of the inverse square law are required to constrain and confront future generations of theories. Encouragingly these constraints can be established at very affordable cost.

GRAVITATIONAL REDSHIFT

Eight years before the full formulation of general relativity, Einstein predicted the gravitational redshift of light. Einstein regarded this as a fundamental prediction of his general theory, but it is now recognised as a test of the equivalence principle. Any theory of gravity that is compatible with the equivalence principle will predict the same gravitational redshift, and in this sense the redshift is really a consequence of the curved space-time.

In principle this test is simple; all that is necessary is to measure the wavelength of a spectral line in the spectrum of a massive star and compare it with the wavelength of the same line measured in the laboratory (which, of course, is on a less massive planet). In practice the test was beyond the capability of early twentieth-century technology. Consider the strong yellow line of sodium seen in the light of street lamps. The wavelength of this line is 589.3 nanometres as measured on Earth. The gravitational force on the surface of Sun is about 3000 times stronger than that on the surface of Earth. In this stronger gravitational field the wavelength of the sodium line will shift by 0.00125 nanometres to the red – in other words it will increase by this amount. This shift could not be measured with the astronomical instruments of 1916, but 10 years later the instrumentation had advanced sufficiently to attempt this measurement. But all measurements between 1927 and 1960 failed to agree with the predicted value. The culprit was the solar surface. The surface of the Sun is not a clean environment like a lamp in the laboratory. The gas on the surface of the Sun is in constant violent motion. The atoms in this gas are in random motion (apart from any ordered motion which may also be imposed on the gas) and the light emitted is Doppler-shifted to the blue or to the red, depending on the motion of the emitting atom relative to us the observers. This random motion increases the width of the line, and any ordered motion shifts the position of the line. These and other effects make it impossible to measure unambiguously the gravitational redshift of photons emitted on the Sun's surface. By the late 1960s the motion and the state of the gas in the solar atmosphere was sufficiently well understood to unravel the gravitational redshift from the shift and the broadening caused by the motion of gas. The measured gravitational redshift agreed with the prediction of general relativity to about 5%. The redshift will be higher on the surface of a star with gravitational attraction higher than that on the surface of the Sun, say a white dwarf. A white dwarf is the terminal stage of evolution of a star. The radius of a white dwarf of solar mass can be 40 to 60 times smaller than that of the Sun. Thus the gravitational field at the surface of a white dwarf can be over 1000 times greater than that on the surface of the Sun and consequently the gravitational redshift will also be higher. Unfortunately, it is difficult to measure the mass and radius of a white dwarf very accurately and the measured gravitational redshift cannot be compared with theoretical prediction with great precision.

In 1960 Robert Pound and Glen Rebka made the first accurate laboratory measurement of gravitational redshift (at the Lyman Laboratory of Physics of Harvard University, Massachusetts, USA). Pound and Rebka measured the change in the frequency of a spectral line when it is emitted at the bottom and then at the top of the 74-feet-high (about 22.5 metres) tower of the Jefferson Physics Laboratory. The fractional decrease in the frequency as the photons rise out of the gravitational field of the Earth to the top of the tower is about 2×10^{-15}. Two assumptions are made in this experiment. First, the mass of photon is assumed to be essentially unchanged between the bottom and the top of the tower, and second, the inertial mass of the photon is assumed to be equal to its gravitational mass or the equivalence principle is implicitly assumed. This experiment is therefore a test of the equivalence principle. (The reference to the mass of a photon may seem surprising, but note that photons have energy and this energy is equivalent to a mass through Einstein's $E = mc^2$.)

To measure the very small change in the frequency of a spectral line it is necessary to have an emitter and a receiver of extremely well defined frequency. Pound and Rebka took advantage of the Mossbauer effect. This was discovered in the late 1950s by Rudolf Mossbauer of the Max Planck Institute in Heidelberg, Germany. Mossbauer showed that an exceptionally narrow gamma ray line could be produced if the line-emitting nucleus is trapped in a crystal matrix. The line is narrow because the surrounding matrix absorbs the perturbing forces that increase the Doppler width of the line. Mossbauer was awarded the Nobel Prize for physics in 1961 for his discovery of this technique. In the citation of the prize the Pound–Rebka experiment was mentioned as one of the many important applications of the Mossbauer effect. Pound and Rebka used the 14.4 kilo-electronvolt (0.086 nanometres) line emitted by radioactive ^{57}Fe (an isotope of iron with 57 protons). They measured the redshift of the line when the source was at the bottom and the detector at the top of the Jefferson tower and a blueshift when the source and detector were interchanged. The measured shifts agreed with theoretical prediction to about 10%. In 1965, Pound repeated this experiment in collaboration with Joseph Snider and improved the agreement with prediction to 1%.

The gravitational redshift has also been tested by comparing synchronised clocks, with one clock on the ground and another in an aircraft. The interpretation of the results of this experiment is not simple because both the gravitational redshift and the time dilation of special relativity affect the travelling clock. During 1971 a travelling clock experiment was undertaken jointly by the Washington University, St Louis and the US Naval Research Laboratory in Washington. A caesium-beam atomic clock was flown (on scheduled passenger flights) due east for 41 hours and due west for 49 hours. The flights made a number of (scheduled) stops *en route* and corrections had to be made for these. The gravitational redshift will depend only on the altitude of the aircraft but the time dilation will depend on the speed of the clock relative to the 'master inertial clock' at rest. Note that the clock in the aircraft is not in an inertial frame: it is not at rest or flown in a straight line with uniform speed. Instead, it moves along a curved path, that is it is continually accelerated. After making all the corrections the results verified the cumulative effect of gravitational redshift and time dilation. The observed redshift agreed with theoretical predictions but with poor accuracy.

A definitive test of gravitational redshift was made in June 1976. This test combined the good points of the Pound–Rebka–Snider experiments and the 'flying clock' experiments. The essential requirement of the gravitational redshift test is a large difference in the strength of the gravitational field between points where clocks or line-emitting atoms can be compared. The Pound–Rebka–Snider experiment was limited because the line-emitting source and the detector could not be separated by much more than the height of the Jefferson Tower. The altitudes at which commercial flights operate limit the 'flying clock' experiment, and complex corrections for scheduled stops and take-offs have to be made. The 1976 experiment, performed jointly by the Smithsonian Astrophysical Observatory and NASA, figuratively combined these two experiments by putting a very accurate clock on a rocket and firing it to an altitude of about 10 000 kilometres or about 1.5 times the radius of the Earth. The accurate clock was a hydrogen-maser frequency standard, and was launched on a Scout-D rocket. During the ascent of the rocket both gravitation and time dilation affect the 'time of the clock'. Initially the time dilation is higher than the gravitational blueshift (the clock is moving into a weaker gravitational field as the rocket ascends) because the rocket is moving at very high speed. As the altitude increases the blueshift gets larger and overtakes time dilation at some height. At the peak of the trajectory the gravitational blueshift is maximum and time dilation is zero as the payload is almost at rest. As the payload begins to drop the time dilation gradually increases and the blueshift decreases. The large amount of data obtained during the June 1976 rocket flight was analysed in the following two years and the observed shifts agreed with predicted shifts, both gravitational and time dilation, to a precision of $7 \times 10^{-3}\%$. The very high accuracy of this test proves that the basic notion of curved space-time was correct.

It does not prove that the theory of gravitation is correct. But the fundamental requirement of Einstein's general theory has been proven to a high level of accuracy.

These tests of gravitational redshift (or blueshift) were performed in the gravitational field of Earth. For completeness it is necessary to conduct similar tests in gravitational fields produced by bodies whose composition is significantly different from that of the Earth. This opportunity was provided when the Voyager 1 and Voyager 2 spacecraft flew past Saturn in 1980. Both spacecraft carried ultrastable crystal oscillators, which provided a precise reference frequency for radio transmissions from the spacecraft to the Earth. The gravitational redshift effect is apparent as a decrease in the frequency of the transmitted radio signal as the craft moves in and out of the gravitational field of Saturn. In the vicinity of Saturn the prediction of gravitational redshift has been confirmed to an accuracy of 1%.

Up to about a decade ago the gravitational redshift and time dilation were just challenging tests for relativistic physicists. However, with the increasing use of the Global Positioning System (GPS) satellites for navigation from the early 1990s these small relativistic effects have a direct bearing on everyday life. The GPS satellites are at an altitude of about 30 000 kilometres and move at about 4000 metres per second relative to an observer on the ground. Relative to this observer, the atomic clock on a satellite ticks faster, because of the gravitational blueshift. For the altitudes of the present GPS satellites the relativistic advance of the satellite-clocks is about 30×10^{-6} seconds per day. This can result in a navigational error of the order of a kilometre, which could be the difference between a cruise missile hitting an enemy artillery position or a marshy swamp.

DEFLECTION OF LIGHT BY GRAVITATIONAL FIELD

The bending of light in the gravitational field was a crucial test of general relativity, primarily because it was capable of experimental verification at the time when it was proposed. Unlike the perihelion shift of Mercury, it was a prediction and not an application to a known problem. This test was undertaken in 1919 by a British expedition to Sobral in North Brazil and to the island of Principe in the Gulf of Guinea off the coast of West Africa. The purpose of the experiment was to discriminate between three possibilities[1]:

(1) The path is uninfluenced by gravitation.
(2) The energy or mass of light is subject to gravitation in the same way as ordinary matter. If the law of gravitation is strictly the Newtonian law, this leads to an apparent displacement of a star close to the Sun's limb amounting to 0.87 arcseconds.
(3) The course of a ray of light is in accordance with Einstein's generalised relativity theory. This leads to an apparent displacement of a star at the limb amounting to 1.75 arcseconds outwards.

In the last two cases the displacement is inversely proportional to the separation between the star and the centre of the Sun. The crucial point here is the factor of two difference between Newton and Einstein. The British expedition measured deflections which were in agreement with Einstein and not with Newton.

In the subsequent 50 years a number of expeditions were mounted to repeat the eclipse observations of 1919. In every expedition there was an improvement in the technology over that of the previous expedition. However, eclipses are notoriously fickle and are almost never observable from well equipped observatories with tried and tested equipment and procedures for accurate astrometry. Also the best equipment and procedure cannot control the weather and eclipses do not wait for clear skies. Even slight cloud can degrade the quality of data by a large and often unknown amount and leave a suspicion about its quality. For all observations made since 1919, the mean value of deflection of a light beam grazing the Sun is 1.82 arcseconds with a spread of 23%. This spread is comparable to errors in some of the observations. While this mean value is sufficient to distinguish between the three alternatives of the 1919 expedition, it is not sufficient to choose between general relativity and the alternative theories which were proposed from the 1960s onwards. The last eclipse expedition was mounted by the University of

Texas to observe the total eclipse of 30 June 1973 from Chinguetti Oasis in Mauritania (West Africa). A supplementary expedition was undertaken in November of that year to make reference observations of the stars. This expedition illustrates all the difficulties that afflict this type of measurements. The expeditions had all the benefits of 1970s technology: superior photographic plates, thermal control in the telescope housing, air-conditioned dark room, de-ionised bottled Evian (yes Evian!) mineral water to develop and wash the plates and many other refinements. But Nature had no respect for these fineries; the accuracy of the data was limited by the adverse sky conditions (the bane of all Earth-bound optical astronomers). The deflection, extrapolated to the solar limb, measured by this expedition was 1.66 arcseconds with an error of 11%. This accuracy was not something you write home about. But by the early 1970s the eclipse method of measuring the relativistic deflection of light was being replaced by more accurate radio astronomy techniques.

Karl Jansky of the Bell Telephone Laboratory made the first tentative radio observations of celestial sources in 1931. During the Second World War, as a result of development of the radar, required for the war effort, there was considerable improvement in the technology required for radio astronomy. After the war these radar scientists turned their attention to radio astronomy, particularly in Britain and in Australia. Radio waves are no different from visible light waves, just of a longer wavelength. Whereas visible light spans the wavelength range from 400 nanometres to 700 nanometres, radio waves span the range from a few tenths of a millimetre to several metres. General relativity predicts the same deflection for visible and radio waves, that is, the theory predicts achromatic deflection. The advantage of radio measurements for the deflection of radiation by a gravitational field is that these measurements can be made with very high precision, accuracy that is impossible in eclipse observations. This higher accuracy is achieved by using radio interferometric observations. In these observations a radio source is observed with telescopes at (at least) two locations. A wave front arrives at the two telescopes at slightly different times. The time difference depends on the angle between the wave front and the line joining the two telescopes, called the baseline. For a given radio frequency (or wavelength) the accuracy with which the angle can be determined increases with the length of the baseline. The accuracy also increases with increasing frequency or decreasing wavelength so that the accuracy is proportional to the ratio of baseline to wavelength, the higher the ratio the greater the accuracy. These days radio interferometer baselines from a few kilometres to intercontinental distances have been used for radio observations. With some of these interferometers a resolution of the order of 0.1 milliarcseconds is possible. Even higher-resolution observations are now possible with interferometers that use the Japanese space-based radio telescope (HALCA), launched in 1992 together with ground-based telescopes.

The accuracy of radio measurements is also limited by the size of the monitored radio source. Most celestial radio sources are radio galaxies and the angular extent of these can be as large as a degree. Fortunately there is a class of powerful radio sources that are point-like, and these are called quasars (quasi-stellar sources). These were once believed to be galactic stellar sources emitting strong radio signals, but it is now established that they are extragalactic objects. The discovery of quasars had motivated the application of general relativity to astrophysical problems (more about this in the next chapter, Hubble & Eddington). But the radio strength and the point-like nature of these sources make them perfect background sources for relativistic deflection experiments. At least two sources, which are close to each other in the sky and which pass close to the Sun, as seen from the Earth, are required for a deflection experiment. The two sources are required because the deflection of the source that is closer to the Sun is measured relative to the one that is further away. This is not unlike the eclipse experiment in which the deflection of the stars close to the Sun, at the time of totality, is measured relative to stars that are further away from the Sun.

Two of the strongest radio quasars 3C273 and 3C279 (3C stands for the third Cambridge (England) radio survey) are separated by 9.5 degrees on the sky. On 8 October each year the sources pass very close to the Sun, as

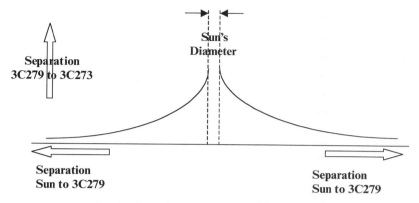

FIGURE 6.2 A schematic representation of the separation between 3C279 and 3C273 as 3C279 is occulted by the Sun on 8 October every year.

seen from the Earth. The quasar 3C279 is occulted by the Sun, but the closest approach of 3C273 to the Sun is about 4 degrees. The separation between the two quasars can be measured as they approach the Sun on 8 October and then as they recede from the Sun. As the quasars approach the Sun the separation between them will appear to increase because of the relativistic deflection of radiation from 3C279. The apparent separation will be maximum when 3C279 is occulted by the Sun, when, of course, it cannot be observed. The separation will begin to decrease as the quasars recede from the Sun. This is shown schematically in Figure 6.2. The deflection of radiation from 3C273 will be very small (of the order of 0.05 arcseconds). By interpolating between the data obtained from the approaching and the receding paths, the deflection, when the radiation from 3C279 grazes the Sun, can be calculated. These observations can be repeated every year, as opposed to eclipse observations, which can only be made sporadically. Irwin Shapiro and his colleagues made the first two successful observations in October 1969. The observations were made from Owen Valley Radio Observatory of the California Institute of Technology and the Goldstone Tracking Station of NASA, also in California, USA. These observations were made with baselines of 1 kilometre and 21.5 kilometres respectively. Deflection angles of 1.77 arcseconds and 1.82 arcseconds respectively were measured for grazing rays. The error in these measurements was of the order of 10%. These observers found that the effects due to the solar corona and the Earth's ionosphere were negligible at the frequencies at which the observations were made. The 10% accuracy, while good enough to confirm agreement with general relativity, is still not good enough to discriminate between general relativity and the competing theories of gravity, in particular the Brans–Dicke theory. In the following three years these measurements were repeated in the United States, Britain and The Netherlands. The baseline of these measurements was similar, of the order of a few kilometres, and the value of the measured angle of deflection was consistent with general relativity. But the error in these measurements was still too big to distinguish general relativity from competing theories. The first Very Long Baseline Interferometer (VLBI) measurements were made in October of 1972. These observations used the 120-foot diameter 'Haystack' and the 60-foot diameter 'Westford' antennas of the Haystack Observatory in Westford, Massachusetts, USA and two 85-foot diameter antennas of the National Radio Astronomy Observatory (NRAO) in Green Bank, West Virginia, USA. The NRAO observatory is 845 kilometres south-west of the Haystack Observatory. The measured deflection for grazing rays was 1.71 arcseconds with an accuracy of 6%. This accuracy is still not good enough to distinguish between general relativity and the Brans–Dicke theory.

The first high-accuracy measurement of deflection of a 'light' beam was made in 1974 and 1975 at the NRAO using a 35-kilometre baseline interferometer. The celestial sources for these measurements were quasars 0116 + 08 (which is occulted by the Sun on 11 April) and 0119 + 11 and 0111 + 02. The first four digits here denote the right ascension, in hours and minutes, of the object. The last two digits denote the declination of the object, the plus sign is for northern objects (a minus sign would denote a southern object). The effects of solar-coronal refraction were removed by simultaneous observations at two frequencies. The measured deflection angles in 1974 and 1975 were 1.78 arcseconds and 1.75 arcseconds with an accuracy of 1 and 0.8% respectively. The agreement between the two measurements made one year apart rules out any underlying systematic error. The result of these measurements favours general relativity over the Brans–Dicke theory.

A factor of five improvement in the accuracy of the grazing deflection angle was achieved with data which were originally obtained for a completely different programme. Between 1980 and 1990 the National Oceanic and Atmospheric Administration's POLARIS and IRIS projects and NASA's Crustal Dynamics Project had accurately measured the positions of a number of celestial radio sources, the aim being to measure regional and global crustal motion, parameters of Earth's rotation and polar motion, and the changes in these rotational rates. These observations were accurate to 10^{-3} arcseconds and covered a large range in the separation distance from the Sun. A network of intercontinental VLBI observatories had obtained these data with baselines between 7000 kilometres and 10000 kilometres. A group at the Laboratory of Geoscience, Maryland, USA, analysed 214 observations of 74 sources. These data had been obtained between 3 degrees and 6 degrees of the Sun, that is deflections between 0.155 arcseconds and 0.077 arcseconds. The data yield a deflection of 1.75 arcseconds at the limb of the Sun with an accuracy of 0.2%. The high quality of these data required corrections for effects that previously had been ignored. Apart from the refractive effects in the solar corona and the ionosphere of the Earth, which were corrected by taking data at two frequencies, effects due to atmospheric refraction, tectonic plate motion, tides, Earth's nutation, ocean and barometric loading and antenna deformation were considered. The effects of the gravitational field of the Earth were also included. The original research programmes for which these data were obtained will continue for at least another decade and the volume of these high quality data will increase substantially. Thus a factor of two to five, if not an order of magnitude, improvement in the quality of this measurement is possible. At this accuracy subtle systematic errors will dominate the measurement of the general relativistic deflection angle.

The launch of the Hipparcos astrometry satellite in 1989 provided an opportunity to accurately measure, at optical wavelengths, the general relativistic value of gravitational light deflection. Analysis of the large amount of accurate data accumulated by Hipparcos suggests that the measured Sun grazing angle is in agreement with Einstein's prediction to an accuracy of 0.5 to 0.7%[2]. This is comparable and compatible with the currently available VLBI observations.

THE 'FOURTH TEST' OR THE TIME DELAY OF RADIO WAVES

In late December 1964, Irwin Shapiro published his 'Fourth test of general relativity' in the scientific journal *Physical Review Letters* – fourth, in relation to the three tests suggested by Einstein. Shapiro proposed that a radio (or light) signal would slow down in a gravitational field. If a radio pulse was sent just grazing the solar limb and then reflected back, the separation in time between the transmission and reception of the pulse would be the time taken by the radio pulse to travel to the reflector and back plus 250 microseconds. To illustrate this, consider a round trip of a radio pulse to Mars when Mars is on the other side of the Sun from the Earth: that is, Mars is at superior conjunction. Mars is 227.9×10^6 kilometres from the Sun and the Earth is 149.6×10^6 kilometres from the Sun. A radio pulse travelling at 3.0×10^5 kilometres per second will make a round trip of 377.5×10^6 kilometres in 2516.7 seconds. Shapiro proposed that if the Sun were close to the line of sight to Mars (actually grazing this line of sight) then this time would be larger by 250 microseconds, an increase of 10×10^{-6}%. It is rather surprising that Einstein had failed to discover this effect. It is possible that Einstein ignored it as the expected delay is so small and totally beyond the measuring capability of early twentieth-century technology. But the same could be said of the gravitational redshift, which Einstein discovered nonetheless. It is just possible that Einstein missed the time delay effect. One ten millionth of a per cent is a small change to measure. Moreover the amount of radio power scattered back to Earth by a planet is about 10^{27} times lower than that transmitted to the planet. But by the late 1950s and early 1960s radio technology had progressed to a level where radio ranging of planets was considered feasible. The driving motive behind the radio ranging measurements was the ambitious plans for space travel by both the USA and the former USSR. Plans for space probes (both orbiters and landers) to the Moon and the nearby planets Venus and Mars were being considered in both countries. To land a probe on a moon or a planet the distances to these bodies had to be known to a few kilometres. Space probes could not be manoeuvred over large distances like cars, and lack of knowledge of accurate distance could make a difference between a heroic landing and an embarrassing miss. The first successful detection of radar echoes, bounced off Venus, was made on 14 September 1959. Radio contacts with Mars and Mercury were made soon afterwards. This opened the new field of planetary radar astronomy. This is a powerful technique; reflected or scattered radar signals carry significant information about planetary surface (topography) and planetary rotation. The surface features of Venus cannot be observed optically as the planet is permanently shrouded in thick cloud; the surface can only be explored by radar. The retrograde motion (rotation in the opposite sense to the orbital motion) of Venus was discovered by radar ranging.

Apart from shedding light on planetary topography, radio ranging can also lead to significant improvements in the determination of planetary orbits. By the early 1960s the distance from Earth to a planet could be measured to an accuracy of almost 1 kilometre. By the mid-1960s radio transmitters and receivers were powerful enough to measure radar echoes off planets at superior conjunction and Shapiro published his paper on the 'fourth test'.

Shapiro and his colleagues also made the first measurement of the time delay. This measurement was made from the Haystack telescope (managed by MIT's Lincoln Laboratory) on 9 November 1966, when Venus was at superior conjunction. The measurement was repeated in January, May and August in 1967 when Mercury was at superior conjunction. To make the time-delay measurement, two sets of measurements of the time of the echo are required: one when the line of sight to a planet is far from the Sun (the reference measurement), and the second when the line of sight is close to the Sun. The difference between the two measurements is the additional delay caused by the gravitational field of the Sun. This methodology is similar to that used to measure the deflection of light (radio waves) in the gravitational field of the Sun. Shapiro and his collaborators measured the expected time delay of 250 microseconds to an accuracy of 20%. The accuracy of this measurement is influenced by the unknown scattering profile of a planet. The average scattering properties of a planet can be determined by echo measurements at inferior conjunction when the signal-to-noise ratio is high. However, this gives no information on the scattering properties of the sub-radar point at superior conjunction. This could be anything, a mountain or a valley, with scattering properties very different from the measured average scattering properties. There is a slight advantage in making this measurement with Venus because its orbit relative to the Earth is well known. Also because of its resonant rotation the sub-radar point at superior conjunction is approximately the same as that at inferior conjunction where the topography can be mapped with better accuracy. Shapiro and his colleagues accumulated data for the next three years from observations of Mercury, Venus and Mars made from the Haystack Observatory and Venus and Mercury observed from the Arecibo Observatory. The accuracy of the time delay measurement was increased to 5%. But this is the limit; further improvement is not possible because of the fundamental limitation imposed by the unknown topography, which cannot be successfully calibrated out of the measurements.

In these measurements of the time delay, the planet really plays a passive role, that of a scatterer or a reflector (mirror). Unfortunately a planet is an imperfect reflector, rather heavily pitted, and it is not possible to map these imperfections very accurately. Clearly, to improve the accuracy of the time delay measurement it is necessary to replace the imperfect mirror with a mirror whose reflecting characteristics are well known and calibrated. This can be done by using a spacecraft as a reflector. A radar signal sent to a spacecraft can be received and amplified by the on-board equipment and reflected back to Earth. The orbit of the spacecraft can be determined by tracking the trajectory accurately. The relativistic time delay can be measured by radar ranging a spacecraft at superior conjunction. An opportunity to make this measurement arose

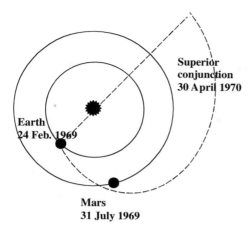

FIGURE 6.3 A schematic representation of the geometry for relativistic time delay measurement with the Mariner 6 spacecraft. The spacecraft launched from Earth rendezvoused with Mars on 31 July and was at superior conjunction on 30 April the following year.

when Mariner 6 and Mariner 7 were launched to fly past Mars to photograph the Martian surface and study the Martian atmosphere. On 31 July 1969 Mariner 6 rendezvoused with Mars when the planet was close to inferior conjunction. The craft flew past Mars and on 30 April 1970 it was at superior conjunction; a schematic of the orbit is shown in Figure 6.3. Mariner 7 was at superior conjunction on 10 May 1970. Radar ranging measurements were started from the Jet Propulsion Laboratory (JPL) in December 1969 and were continued for almost twelve months. The heaviest concentration of observations was around the time of superior conjunction of the two crafts when measurements were made almost every day. At superior conjunction the distance to Mariner 6 was 402.25×10^6 kilometres and that to Mariner 7 was 385.56×10^6 kilometres. The corresponding Newtonian time delays are 1340.8 seconds and 1285.2 seconds respectively. Because of the tilt of the orbits of the two Mariner spacecraft the line of sight to the crafts did not graze the limb of the Sun. The line of sight to Mariner 6 was about 1 degree north of the Sun and that to Mariner 7 was about 1.5 degrees north of the Sun. Thus the closest that the line of sight to Mariner 6 came to the Sun was 3.5 solar radii and the Shapiro time delay was 200 microseconds. The corresponding distance and time delay for Mariner 7 were 5.9 solar radii and 180 microseconds respectively. The JPL group measured these delays with an accuracy of 3%. This was a considerable improvement over the radar ranging measurements of planets but it was not good enough to distinguish between general relativity and alternative theories like the Brans–Dicke theory.

It was, however, realised that the radar ranging of spacecraft had its own limitations. The crafts were light and were constantly buffeted by the solar wind, a stream of electrons, protons and ions flowing out from the Sun. It is this wind which causes the long tail of comets. The wind pressure has two components, a constant component caused by the steady wind and a varying component. The constant component is well mapped but the varying component depends on the unpredictable changes taking place in the solar atmosphere and cannot be mapped. The attitude control system of the spacecraft also makes small but constant adjustment to the spacecraft to keep the solar panels, that power the craft, oriented to the Sun and the communication antennae pointed to the Earth. The Sun and the attitude control together introduce small random changes in the orbit of the craft and this can create a significant discrepancy between the predicted orbit and the real orbit. This translates into an error in the distance between Earth and the spacecraft or an error in the radar echo timing. The radar ranging of a planet does not suffer from these problems because a planet is massive and is not buffeted by the solar wind and there is no attitude control system.

To improve the accuracy of radar echo timing it was necessary to take advantage of the strong points of the two methods: anchor the spacecraft on a planet. This opportunity presented itself during the Viking lander missions to Mars. The primary objective of the Viking lander missions was to transmit a close-up view of the Martian surface, analyse the atmosphere and the surface composition and look for signs of life on Mars. But this was also an excellent opportunity for relativists to make accurate measurements of the radar echo time delay. The first Viking spacecraft reached Mars by mid-June in 1976 and Lander 1 descended to the plain of Chryse on 20 July. The second Viking spacecraft also arrived at Mars in the summer of 1976 and on September 3, Lander 2 was dropped on the region called Utopia Planitia. Early in September the relativists at

JPL and MIT started ranging measurements, as the superior conjunction was on 26 November. Both landers had S-band and X-band transponders corresponding to frequencies of 2.3 gigahertz and 8.4 gigahertz respectively, and this allowed a very accurate correction to be made for the delay caused by the corona of the Sun. In all previous delay measurements this effect had been corrected with the aid of models of the solar corona. Ranging measurements were made from about September 1976 to September 1977 and included the period of superior conjunction. The analysis of data achieved an accuracy of 0.2% in the measured Shapiro time delay. This high accuracy result is in excellent agreement with general relativity. The result also severely restricts the validity of other theories of gravitation like the Brans–Dicke theory. Further improvements in this result will be possible with future planetary landers.

EQUIVALENCE PRINCIPLE

The principle of equivalence – the hypothesis of complete physical equivalence between a gravitational field and an accelerated reference frame – has played a central role in the theories of gravitation. Almost 400 years ago Galileo is supposed to have performed his famous 'Leaning Tower of Pisa' experiment to show that, but for air resistance, spheres of wood and lead fall with the same acceleration. Newton showed theoretically and experimentally that the gravitational mass of a body was proportional to its inertial mass. This demonstration of the principle of equivalence was to be the cornerstone of his theory of mechanics. Newton's equivalence principle (although Newton did not name it or put it this way) can be stated thus[3]:

> If an uncharged body is placed at an initial event in space-time and given an initial velocity there, then its subsequent trajectory will be independent of its structure and composition.

Here 'uncharged' means electrically neutral. This principle is valid for bodies that have negligible self-gravitational energy – that is, laboratory-size bodies. Einstein, in developing the general theory, formulated the Einstein Equivalence Principle (EEP)[3]:

Newton's equivalence principle is valid, and the outcome of any non-gravitational test experiment is independent of the velocity of the (freely falling) apparatus and also independent of where and when in the universe it is performed.

Stated thus the EEP incorporates both the local Lorentz invariance and the local position invariance. The EEP is at the heart of Einstein's gravitation theory (actually at the heart of a family of gravitation theories) because what it is saying is that if EEP is valid then gravitation is a consequence of the curved space-time. Thus tests to verify EEP are fundamental to check the validity of any theory which describes gravitation in terms of space-time curvature. Tests of increasing accuracy, as described below, have been performed over the past hundred years and future tests of even higher accuracy are planned.

In the mid-1960s Kenneth Nordtvedt of Montana State University, Bozeman, USA, reconsidered the EEP. He asked a simple question: suppose the 'Leaning Tower of Pisa' experiment was repeated but instead of dropping small spheres of wood and lead you were to drop the Sun and a white dwarf. Would they fall with equal acceleration? This question is worth asking because there is a fundamental difference between laboratory-size objects and massive bodies like the Moon and the Sun. A laboratory-size object is held together by electromagnetic and nuclear forces, and the gravitational attraction between the constituents of the object is very small (that is, the self-gravity of the body is small). But a massive body is held together by its own gravity (the self-gravity is very large). Put differently, the gravitational self-energy of a laboratory-size object is considerably smaller than the rest energy of the object. For a massive body the self-energy is a larger fraction of the rest energy. The ratio of gravitational self-energy to rest-energy of a laboratory-size object is about 10^{-27}. This ratio for the Moon is about 10^{-11}, for the Earth it is about 10^{-10} and that for the Sun it is 10^{-6}.

The universality of gravitation, or its independence from the composition of a body and from the forces holding the body together, allows gravity to be interpreted as a curvature of space-time. But the force holding a massive body together is the same force that attracts two bodies together. Suppose there is some nonlinear gravity–gravity interaction between the internal gravitational field of a body and the external gravitational field. Massive bodies of different mass, or different internal gravitational field, may fall with different acceleration in an external gravitational field. Nordtvedt showed that in general relativity, the acceleration of the two massive bodies would be same; in other words, the acceleration is independent of the internal gravitational field. This can be generalised as the Strong Equivalence Principle (SEP):

> If an uncharged body is placed at an initial event in space-time and given an initial velocity there, then its subsequent trajectory will be independent of its internal energy.

The secondary clauses of the local Lorentz invariance and the local position invariance should be included in the SEP. Thus, according to general relativity, the acceleration of an aluminium ball and a white dwarf will be the same in an external gravitational field. It is interesting to recall that Newton had considered the validity of this principle to be self-evident, as he asserted in the *Principia* (see Chapter 4, Newton).

The pendulum experiments of Newton proved that the inertial and gravitational masses are equal to an accuracy of 1 part in 10^3 or the accelerations of bodies of different composition are equal to this accuracy. In 1832 the celebrated German astronomer Friedrich Wilhelm Bessel (1784–1846) improved the accuracy of this experiment by about a factor of 100. Bessel is better known for making the first measurement of the distance to a star other than the Sun, by measuring the annual parallax of the star called 61 Cygni. He thus confirmed the enormous distances to stars which Copernicus, Tycho Brahe and Galileo had suspected. In the late nineteenth century Vásárosnaményi Báro Eötvös Loránd, usually Germanised to Baron Roland von Eötvös (1848–1919), started a series of experiments which significantly improved the accuracy of the test of the equivalence principle. Eötvös (pronounced *ut-vush*) was born in Budapest in Hungary on 27 July 1848. Son of a distinguished Hungarian statesman, novelist and political philosopher, Eötvös studied law in Hungary. But in 1869, at the age

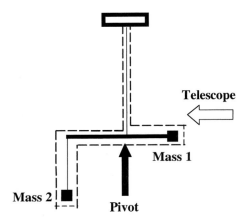

FIGURE 6.4 A schematic diagram of Eötvös's apparatus. The deflection of the horizontal beam could be observed by viewing a calibrated scale with a telescope. The balance was enclosed and the whole apparatus could be rotated round the pivot.

of 21, he went to the University of Heidelberg to study physics. At Heidelberg, Eötvös came under the influence of distinguished men of science of nineteenth-century Germany, physicists Hermann von Helmholtz and Gustav Kirchhoff and the chemist Robert Bunsen. Eötvös obtained a doctorate in theoretical optics and at 24 was appointed a professor in Budapest. In his early years Eötvös was involved with experiments on molecular phenomena, topics which were at the forefront of research in physics and chemistry at the time. Like his father, Eötvös also served in the Hungarian government and introduced a number of reforms, which directly contributed to the rapid rise of Hungarian science after the First World War. Sometime in his late thirties, he turned his attention to gravitation and an accurate test of the equivalence principle. He used a version of the torsion balance used by Cavendish to measure the value of the universal constant of gravitation. The instrument he used is shown schematically in Figure 6.4. It consisted of a light horizontal beam, 40 centimetres long, suspended by a thin platinum–iridium wire. Attached to the ends of the beam were two masses, one suspended 20 centimetres lower than the other. This configuration of the torsion balance was unnecessarily complicated for tests of the equivalence principle. In a digression to geophysics, Eötvös used a modified version of his apparatus for measuring the distribution of mass in mountain ranges.

The principle of Eötvös' experiment can be explained thus. Two forces act on an object on Earth's surface; a force towards the centre of the Earth (this force is represented by the object's gravitational mass) and a force attempting to move the object along a tangent to Earth's orbit (the resistance of the object to move along the tangent represents its inertial mass, these days called simply 'mass'). The force attempting to move the object along the tangent is the centripetal force due to Earth's rotation. The net force is along a line at a slight angle to the vertical. If the ratio of inertial to gravitational mass is different for bodies of different composition then this angle to the vertical will be different for these bodies. In the Eötvös experiment this difference will twist the horizontal beam. Eötvös aligned the beam east–west and noted the rest position of the beam. He then rotated the whole apparatus through 180 degrees, so that any difference in the 'inertial pull' would have caused the beam to rotate in the opposite direction and come to rest in a different position. If there was no difference in the inertial force on bodies of different material there would be no torque, and no difference in the orientation of the beam in the two configurations of the apparatus. A null result would indicate that the inertial mass of a body was independent of the composition of the body. Eötvös and his collaborators performed their first test of the equivalence principle in 1898 using platinum for one mass and copper, water, aluminium and so on for the other mass. Various corrections for stray gravitational forces exerted by experimenters themselves (the Baron was no lightweight!) were necessary. The experiment was repeated in 1908 and the equality of inertial and gravitational masses was proved to a remarkable accuracy of 1 part in 10^9. A paper describing this work won Eötvös, in 1909, the Benecke Prize of the University of Göttingen. For some reason the paper was not published until 1922, three years after the Baron's death. These experiments pre-dated Einstein's work on gravitation, but he was not aware of the results. But if the Baron had not obtained a null result it would have been noted and commented on by other scientists and would certainly have influenced Einstein's work. The results of the Eötvös experiments were brought to Einstein's attention around 1912 and he referred to these extensively in his subsequent publications.

These results were considered to be definitive for the next 50 years and no attempt was made to repeat the experiments or improve their accuracy. However, with the reawakened interest in general relativity and the flowering of alternative theories of gravity in the late 1950s and 1960s there was renewed interest in the equivalence principle. There was also a deeper understanding of the crucial rôle of the equivalence principle in the general theory of relativity and the family of theories in which the gravitational attraction is a consequence of curved spacetime. Two experiments were undertaken in the late 1960s and early 1970s. One experiment, led by Dicke, was at Princeton University and the other, led by Vladimir Braginsky, at the Moscow State University in the former USSR. Both experiments were torsion balance experiments, similar to that of Baron Eötvös, but with a number of modifications and improvements. Both groups had the advantages of mid-twentieth-century technology, excellent fibres for supporting the horizontal beam and the weights, good vacuum systems to eliminate the effects of air currents, state-of-the-art temperature control, and sophisticated electro-optical systems for measuring the deflection of the beam. The experimenters also removed the human observer to eliminate the personal bias introduced by observers. These experiments replaced the inertial centrifugal force of the Earth's rotation by that due to the Earth's orbit around the Sun. This had two significant advantages: the unsatisfactory procedure of bodily rotating the apparatus through 180 degrees was avoided as the Earth did the rotation and the observations were made continuously (because the apparatus is rotating continuously round the Sun) and not in two discrete steps as Eötvös had done.

To understand this experiment, consider the schematic representation in Figure 6.5. The aim of the experiment is to measure the relative attraction of two bodies by the Sun. Consider the layout of the two masses at say 6 a.m. And suppose the line joining the two weights is perpendicular to the line-of-sight to the Sun. If there is a difference in the attraction of the two masses by the Sun then a deflection will be

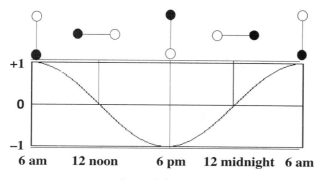

FIGURE 6.5 Relative deflection of the torsion balance in the Princeton (or Moscow) test of the equivalence principle. If there is a difference in the force acting on the two dissimilar bodies then in a 24-hour period the swing of the torsion balance will define a curve similar to that shown. If there is no difference in attraction, the balance will not swing. (Adapted from C. M. Wills, *Was Einstein Right?*)

registered. At noon the two masses will be along the same line to the Sun and there will be no deflection. At 6 o'clock in the evening the two masses will have rotated through 180 degrees and there will be deflection but in the opposite direction to that registered at 6 a.m. At midnight the masses will again be on the same line to the Sun and there will be no deflection. The cycle will repeat after that. The deflections of the beam will trace out a curve similar to that shown in Figure 6.5. If there is no difference in the attraction of the two masses then the curve in Figure 6.5 will be a straight line at zero deflection. Note that the alignment of the two bodies *relative to Earth* does not change in this experiment and the effect due to the Earth's rotation stays fixed. The Princeton experiment had weights of copper and lead (actually lead chloride, which can be highly refined by repeated crystallisation). This choice was governed by a desire to have weights with very different composition of protons and neutrons: copper has 29 protons and 34 neutrons, in lead the ratio is 125 to 82. The Princeton experiment proved the equivalence principle to an accuracy of 1 part in 10^{11} and the similar Moscow experiment achieved an accuracy of 1 part in 10^{12}.

Eric Adelberger and his colleagues at University of Washington have performed more recent torsion balance experiments. They have established the equivalence principle to an accuracy of 2.4 parts in 10^{12}. Ramnath Cowsik and his colleagues hope to establish the equivalence principle to a precision exceeding 1 part in 10^{13} with a torsion balance experiment under construction at the Indian Institute of Astrophysics in Bangalore.

Because the equivalence principle is of fundamental importance to theories of gravitation, experimental relativists are constantly looking for ways of improving the accuracy of the tests and new methods to test the principle. The next STEP (Satellite Test of the (Weak) Equivalence Principle) is to repeat Galileo's free-fall experiment in space. The STEP experiment has been proposed by research groups in the USA and Europe and has been adopted by both NASA and the European Space Agency (ESA). The principle of the experiment is as follows. In a satellite around the Earth, four pairs of two cylindrical test masses (of different composition) are mounted concentrically and supported by magnetic bearings, which leave them free to move along their common axis but not sideways. In orbit the test masses are maintained in a fixed position relative to the line from the satellite to Earth (see Figure 6.6). As the satellite moves in its orbit, each test mass feels simultaneously the gravitational acceleration of the Earth and the inertial centripetal acceleration due to the satellite's orbital motion. If the accelerations differ for different materials, the two test bodies will oscillate with respect to each other once per orbit. The STEP experiment has two advantages over the experiment performed by Galileo. Firstly, the 'STEP Tower of Pisa' is almost 7000 kilometres high and not just 50 metres high, and secondly, the separation of the falling weights can be measured with a precision which is almost a trillion (10^{12}) times higher than was possible in Galileo's experiment. For STEP the 'tower' is the radius of the orbit of the satellite round Earth and the ultra-precise measurement is possible with the cryogenic technology of the late twentieth century. The STEP experiment will attempt to measure the separation of the test masses to an accuracy of about 10^{-12} metres – less

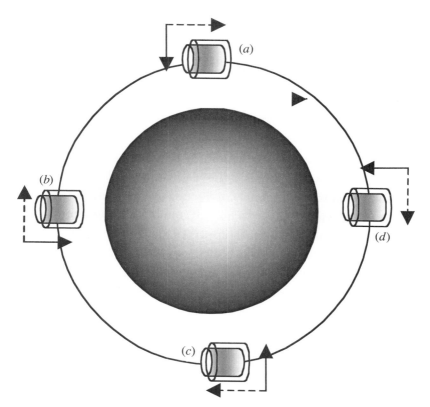

FIGURE 6.6 A schematic diagram of the orbit of the STEP satellite. The gravitational attraction is denoted by the full arrow and the centripetal force by the dotted arrow. At positions (a) and (c) in the orbit, the gravitational force acts along a line perpendicular to the axis of the two masses (of different materials) and therefore cannot cause relative motion between them, but the centripetal force acts along the axis of the mass. If this force is different for the two masses then they will move relative to each other. At positions (b) and (d) the centripetal force will cause no relative motion as it acts along the line perpendicular to the axis of the masses but the gravitational force acts along the axis and again, if there is a difference in the gravitational attraction of the two masses, they will move relative to each other. If there is no difference in the gravitational and centripetal forces acting on the two dissimilar masses, there will be no relative change in their separation.

than about 1% of the diameter of an atom. This high accuracy is possible with a SQUID (Superconducting QUantum Interference Device), a device for measuring tiny changes in the magnetic field. It is projected that the STEP experiment will test the equivalence principle to an accuracy of 1 part in 10^{18}.

This brief description of STEP conceals a spacecraft of extraordinary complexity. The experiment is immersed in a dewar containing 180 litres of liquid helium at a temperature of 1.8 degrees above absolute zero. This low temperature is required to operate the SQUID devices and also to reduce the thermal noise in the test masses and minimise the residual gas pressure. The whole experiment has to stay at this low temperature for the operational life of the project, about eight months. Thus the tank of helium has to be designed such that the evaporating helium is used to cool the tank and thus reduce the loss of liquid helium. Special precautions are necessary in the design of the tank and the test masses to reduce the effects of helium tides raised by the Earth. The STEP satellite will be at an altitude of about 500 kilometres and although the atmospheric drag on the satellite at this altitude is small, it is huge compared with the accuracy aimed at by experimenters. It is therefore necessary to compensate for this drag, and this compensation is provided by controlled release of the helium escaping from the dewar. Although the complexities of the STEP satellite are daunting they are within the capabilities of present technology. The crucial question is – is STEP affordable? If sufficient funding is made available the STEP satellite may be launched in the first decade of this century.

The Strong Equivalence Principle (SEP) cannot be tested in a laboratory or in satellite-based experiments because the gravitational binding energy of any test mass will be too small. Nordtvedt proposed that the SEP could be tested by monitoring the acceleration of the Earth and the Moon towards the Sun. If SEP were violated (the Nordtvedt effect) then the Moon's orbit would be elongated in the direction of Sun by about 1.3 metres. The detection of this elongation has been possible since 1969 when the Apollo 11 astronaut placed retro-reflectors on the Moon for the lunar laser ranging (LLR) experiment. In 1976 Shapiro and his group at MIT published the results of four years of observations and showed that there was no elongation of the Moon's orbit, and the SEP was proved to an accuracy of 7 parts in 10^{12}. Analysis of LLR data obtained from 1969 to 1999 has established that the Earth and the Moon, bodies of different composition and different internal gravitational field, fall towards the Sun at rates that are equal to 1 part in 10^{13}. The Earth–Moon system is however a weak field system; the gravitational self-energy of both the Earth and the Moon is relatively small. Tests of the SEP in a strong field regime are possible with the binary pulsars. The gravitational self-energy of a neutron star is high, the ratio of self-energy to rest energy being about 0.2. These experiments are still in their infancy but the violation of SEP appears to be excluded at the 0.5 per cent level in these strong field systems. Both WEP and SEP have now been tested to a reasonably high accuracy. However, the equivalence principle is so crucial to the theories of gravitation and to general relativity that even higher-accuracy tests will be made in future. Also, theories such as the superstring theory (see Chapter 8, Planck) which aim to unify the four forces of nature suggest the violation of the equivalence principle at less than the 10^{-12} level. Equivalence principle tests are perhaps the most sensitive low energy probes for this new physics.

GRAVITOMAGNETISM

In the early 1960s Leonard I. Schiff and his colleagues William M. Fairbank and Robert H. Cannon at Stanford University, Stanford, USA, revived interest in gravitomagnetism. Schiff and his colleagues were interested in measuring relativistic effects on a gyroscope in orbit round the Earth. At about this time George E. Pugh, working independently at the Pentagon (part of the huge military establishment of the USA) had also considered relativistic effects on a spinning gyroscope. The first gyroscope (which means 'to see circulation'), a gimballed balanced flywheel, was developed by the French physicist Jean-Bernard-Léon Foucault (1819–1868) in 1851 to demonstrate the rotation of Earth. The axis of a spinning gyroscope always points in the same direction relative to an inertial frame or relative to the distant stars. This is why gyroscopes are of such crucial importance in navigation. With his gyroscope and a 67-metre-long pendulum (the 'Foucault pendulum') Foucault measured Earth's rotation to an accuracy of 0.17%. For this measurement he was awarded the Coply Medal of the Royal Society of London. The gyroscope was ignored for the next 47 years until in 1898, the Austrian engineer Ludwig Obry was able to translate the gyroscope's directional information into steering a torpedo. Despite its inauspicious beginning the gyroscope has proved enormously useful in navigation.

According to general relativity, in curved space near a massive body a gyroscope will not point in a fixed direction. The axis of the gyroscope will precess. Two distinct general relativistic effects are involved. The first, called the 'geodetic effect', is a consequence of the curved space-time near the massive body. In a flat space-time the direction of the spin axis stays fixed, in other words the direction at any instant is parallel to the direction at any previous instant. But in a curved space-time, a locally parallel axis will not be parallel globally; that is, the direction at any instant will be parallel to the adjacent axis but will not be parallel to the axis at any other time. Thus upon completing a closed path, the gyroscope axis will point in a direction different from the direction at the start. The Dutch physicist Willem de Sitter first calculated this geodetic effect. In 1916, he showed that the relativistic perturbation due to the gravitational field of the Earth and the Sun would cause the Earth–Moon system to precess at a rate of about 19.2×10^{-3} arcseconds per year. Eddington and others soon pointed out that the Earth–Moon system was a gyroscope and the de Sitter effect was the precession of a gyroscope axis. Unfortunately the effect was too small to be measured in 1916. This effect – the precession of the local inertial frame with respect to distant inertial frame – has now been measured with radio interferometers and lunar laser ranging, with a precision of three parts in 1000. Measurements of far greater accuracy will be possible with a gyroscope in orbit round the Earth.

The second effect is called 'frame-dragging' or gravitomagnetic effect. Two German physicists, J. Lense and Hans Thirring, proposed this in 1918. The proposal was that a moving or a rotating body would produce a gravitational field analogous to the magnetic field produced by a moving or rotating electric charge (see Chapter 5, Einstein). This is the only effect that tells us something about the inertial properties of the space-time. Lense and Thirring had considered the rotation of a body like the Sun and the effect of the resulting frame-dragging on the orbits of the planets. Unfortunately the effect is so small that the possibility of detecting it was not even considered in 1918. But the swirls in the gravitational field produced by a rotating black hole may actually have been observed. In 1977 Roger Blandford and Roman Znajek, working in Cambridge, England, proposed that magnetic fields threaded through these swirls could accelerate charged particles. These accelerated charged particles have been seen as huge jets, which extend over millions of light-years from the nuclei of active galaxies.

In 1984 it was suggested that frame-dragging in Earth's gravitational field could be measure using the satellites of LAGEOS (LAser GEOdynamics Satellite) type. These are high altitude, high mass-to-area ratio, spherical satellites. The 0.6-metre diameter LAGEOS has a brass core with an aluminium shell and weighs 411 kilograms; it carries no instrumentation but is covered with 426 retro-reflectors. The geometry and the material of the satellite are selected to make it as heavy as possible to minimise the effects of all nongravitational forces but accommodate as many retro-reflectors as possible. By measuring the time between transmission of a laser pulse and the reception of the signal reflected off the satellite the distance between the satellite and the ground-station can measured very precisely. These distances can be used to measure the separation of ground stations. The high mass-to-area ratio and the precise stable (attitude-independent) geometry of the satellite, together with the extremely regular orbits, make LAGEOS a very precise position reference. Also the very long life of the satellite means that data can be accumulated over very long periods to improve the accuracy of the positions. At present the distance between ground-stations can be measured within 1–3 centimetres. Long-term data sets can be used to monitor the motion of the Earth's tectonic plates, measure the wobble of the Earth's axis of rotation, and better determine the length of an Earth day. Also, the axis perpendicular to the orbit can be identified as a gyroscope axis. This axis will precess because of the gravitomagnetic effect.

LAGEOS was launched in 1976 by NASA. This satellite has a semi-major axis of 12 270 kilometres and an inclination of 109.94 degrees. The accuracy of tracking its orbit is about one centimetre over 5900 kilometres. The frame-dragging precession of the orbit of this satellite is 31×10^{-3} arcsecond per year. The measured positions of the satellite are accurate enough to detect this. But this precession cannot be extracted from the observed precession of the orbit. This is because the classical precession of the orbit (due to the nonspherical shape of the Earth and inhomogeneous distribution of mass in the Earth) is about 126 degrees per year. This precession cannot be determined (theoretically) with an accuracy of a few parts in 1000 because the true figure of the Earth and the true distribution of mass in the Earth have not been precisely determined. This inaccuracy of the classical precession masks the frame-dragging precession. But the frame-dragging precession can be determined from the precession of orbits of two similar satellites provided the orbits are identical but have different inclinations. In this case the classical precession of the two orbits is the same but the frame-dragging precession can be reversed by a suitable choice of the orbits. Thus the classical precession can be calibrated out. This has been achieved by an Italian group of physicists. In 1992, LAGEOS II, built by Agenzia Apaziale Italiana, was put in an orbit similar to that of LAGEOS but with an inclination of 52.6 degrees. By comparing the orbital precessions of LAGEOS and LAGEOS II the frame-dragging precession in Earth's gravitational field has been measured to an accuracy of 20%.

In January 1961 Schiff and Fairbank decided to employ the gyroscope in the service of fundamental science, namely to understand the nature of space-time. They proposed to NASA a satellite experiment to test general relativity by measuring in space, the precession of a gyroscope axis. Although some research funding was provided, it was another 30 years before the actual satellite programme was funded – this is the proposed Gravity Probe B or GP-B satellite. Conceptually the experiment is very simple but technically extraordinarily challenging. The proposed instrument consists of four gyroscopes, a telescope and a drag-free proof mass enclosed in an evacuated cylinder about 25 centimetres in diameter and 244 centimetre long. This assembly will be immersed in a cylindrical dewar 306 centimetres long and 213 centimetres in diameter. The dewar will be filled with liquid helium and will maintain the experimental assembly at 1.8 degrees. The planned satellite will be in a polar orbit at an altitude of 650 kilometres with a predicted operational life of approximately two years. The telescope on the satellite, whose axis will be aligned with the axis of the satellite, will be pointed towards the star HR 2703 to provide a precise reference axis for the instrument. This star was chosen because it is a 'radio-loud' star and its position and absolute proper motion, with respect to the background quasars, can be measured vary accurately by VLBI (Very-Long-Baseline Interferometry) techniques. The pointing of the telescope is planned to be accurate to $\pm 20 \times 10^{-3}$ arcseconds. The four gyroscopes will be in a straight line also on the axis of the spacecraft with their spin axis parallel to the line of sight to the guide star. Two gyroscopes will be spun clockwise and the other two anticlockwise. At the altitude of the spacecraft the geodetic precession of the gyroscope axis is 6.6 arcseconds per year and the gravitomagnetic precession is 42×10^{-3} arcseconds per year. In addition there is a precession of 19.2×10^{-3} arcseconds per year due to the gravitational field of the Sun. This is the precession de Sitter had computed for the Earth–Moon system. There is also a precession of -7×10^{-3} arcseconds per year due to the oblate Earth.

The satellite builders have to overcome a number of challenging technical problems. To mention but a few:

- The 'guts' of each gyroscope in this experiment is a 2-centimetre radius quartz sphere. This has to be spherical and homogeneous to a few parts in a million to minimise differential torque on the gyroscope.
- The residual gravitational force on the gyroscope has to be reduced by a factor of a billion below that at the Earth's surface. This demands a drag-free satellite.
- The change in the direction of the spin axis of the gyros has to be measured without violating the spherical figure and the homogeneity of the quartz sphere. This is achieved by coating the sphere with a superconducting film. Any change in the direction of the spin axis will induce a current in a superconducting coil surrounding the gyroscope. But for this method to work the gyroscope has to be magnetically shielded to reduce the ambient magnetic field to a level that is a million times lower than the ambient field of the Earth.
- The direction to the reference star has to be monitored to better than 0.1 arcsecond per year and a sophisticated optical system is required to achieve this.

Most of these problems have been solved or the possible solutions are known and are waiting to be implemented. With luck GP-B will be launched in the first decade of this century.

GRAVITATIONAL WAVES

The observations of the binary pulsar by Taylor and Hulse have confirmed that gravitational waves, postulated by Einstein in 1916, do exist. These observations have verified the propagation of gravity with the speed of light (to an accuracy of one part in 1000) and confirmed that the gravitational waves carry energy. But the actual detection of the waves has not been achieved yet. The principle behind experiments to detect gravitational waves is rather simple. A gravitational wave moving past a body will vibrate the body and to detect the wave it is only necessary to observe these vibrations. However, two test-bodies are required to detect a wave just as a single cork floating on a pond or on the ocean will not reveal the passage of a wave. But if there is a second cork floating near it then the passage of a wave will be revealed by the change in the separation between the two corks. This principle can also be used to detect gravitational waves. There is just one problem: the separation the gravitational waves produce is extremely small. Even the strongest wave will move the particles in a test mass. 1 kilometre long only by about 10^{-18} metres, or about one-hundredth of the diameter of a proton! To detect this wave a detector has to be able to measure the change in the shape of a test mass of just 1 part in 10^{21}. The gravitational-wave-induced movement of the particles in a test mass is smaller than the thermal motion of atoms in the test mass. This has not deterred experimental relativists from attempting to detect gravitational waves.

The passage of a gravitational wave will appear as a periodic tidal force, which will squeeze and stretch a test mass in the direction perpendicular to the direction of propagation of the wave. Joseph Weber set out to detect this. Born in Paterson, New Jersey, USA, in 1919, Weber was educated in the US Naval Academy and the Catholic University of America, and joined the faculty of University of Maryland in 1948. In the late 1950s Weber began work on the problem of detecting the tidal forces produced by gravitational waves. By 1965 he had worked out the complexities involved in detecting these waves and put together a simple detector. The detector, now called the Weber Bar, was an aluminium (because it is cheap) bar about 2 metres long and about 0.5 metres in diameter (Figure 6.7(a)). Waves travelling at a right angle to the long axis of the bar will stretch and compress the bar along this axis (Figure 6.7(b)) and the bar will oscillate at a frequency characterised by the size and shape of the bar. If an incoming gravitational wave contains frequencies close to this frequency then the bar will resonate and amplify the vibrations. Piezoelectric sensors attached to the bar convert these tiny displacements to electric signals, which can be detected and analysed. Weber chose the size of his bar, apart from cost considerations, to match the expected frequency of gravitational waves from some astronomical sources. A likely source of gravitational waves is

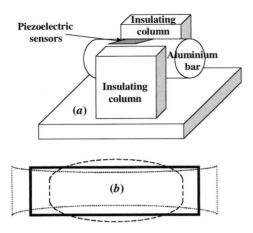

FIGURE 6.7 Weber Bar gravitational wave detector (a). A gravitational wave travelling at right angles to the long axis of the bar will alternately compress and stretch the bar as shown, in a highly exaggerated representation, in (b). There are no forces acting along the direction of propagation of the wave.

coalescing binary stars. Models of coalescing binaries suggest that as the stars spiral towards their common centre of mass, the orbital frequency will increase to about 1 kilohertz. Weber guessed that the emitted gravitational waves would include this frequency. Gravitational waves of similar frequency are also expected from other sources such as colliding black holes and exploding supernovae. Weber's bar had a resonant frequency around 1 kilohertz and a sensitivity of 1 part in 10^{15}. In 1968 Weber stunned the scientific community by announcing that he had simultaneously detected signals in two detectors placed about 1000 kilometres apart, one in Maryland and another at the Argonne National Laboratory near Chicago. The choice of two detectors was deliberate. Apart from the gravitational waves the bar also responds to signals by seismic disturbances (cars, trucks and people moving close to the detector) and random thermal motions of atoms in the bar. But in two detectors, separated by a large distance, these secondary signals will be very different. In such separated detectors only the signals due to gravitational waves will be correlated. This was the startling news in Weber's announcement; he was detecting correlated signals. The correlation suggested a single source for the signals and gravitational waves appeared to be the most likely source. More remarkably, in 1970 Weber announced that the rate of events increased when the detectors were oriented perpendicular to the direction to the centre of the Galaxy. It was tempting to speculate that the Galactic centre was emitting these waves. However, there were two disconcerting aspects of this detection. One was that the strength and the rate of signals were higher than the predictions of theorists. This was not so serious as theorists are often wrong. The really disturbing aspect was that independent research groups, who, between 1970 and 1975, had built detectors with sensitivities comparable or higher than those achieved by Weber, had failed to detect any signal apart from the inevitable noise. Weber's coordinated signals are no longer regarded as being caused by gravitational waves but there is no really good explanation for the coincidences either.

Modern versions of Weber Bar detectors are cooled to liquid helium temperature to reduce the thermal noise. These detectors can have a sensitivity of 1 part in 10^{18}. Such detectors are being built in the US, Switzerland and Italy. Similar experiments but with spherical detectors are also being considered, in the US, Brazil and the Netherlands. Spherical detectors have an advantage in that they are sensitive to waves from all directions.

Another way to detect gravitational waves is with the Michelson interferometer (Figure 5.4) with the light source replaced by a laser. A gravitational wave passing perpendicular to the plane of the interferometer would periodically stretch and shrink the two arms of the interferometer. This change in the length of the two arms would result in a shift in the interference fringes at the detector. Decades of development have gone into developing the principle of Michelson's interferometer into a gravitational wave detector. The detector is operated at high vacuum to prevent pressure fluctuations, which could affect the path of light in different parts of the instrument. To eliminate ground vibration, the two test masses in the two arms of the interferometer are suspended from wires, which are attached to vibration-absorbing supports. Mirrors for reflecting the laser beam are attached to these test masses. The material of the mirror supports and the shapes of the mirrors are chosen to dampen the natural frequencies of the components. However, an intrinsic source of noise in this detector is a quantum effect called the photon shot noise. A high light intensity is needed to diminish the effects of quantum fluctuations, but the required power is not available in a continuous beam. A number of clever tricks have been invented to overcome this problem. Developments in continuous solid state lasers have also made higher powers available. But this higher power brings with it new problems. For example, the power absorbed by the beam splitter could heat it enough to change its refractive index and destroy the interference effect. The ingenuity of physicists has overcome a number of such problems, and detectors with a sensitivity of 1 part in 10^{22} are being built. An interferometer detector of gravitational waves has two major advantages over the Weber Bar. Firstly, an interferometer detector is sensitive to almost all frequencies of gravitational waves; in other words, it has a wide bandwidth. Secondly, the sensitivity of an interferometer detector can be increased considerably by placing the reflecting mirrors a long way from the beam splitter or equivalently, by the beams in the two arms reflecting a large number of times.

Several programs to build interferometric gravitational wave detectors are under way in many countries. The biggest, LIGO (Laser Interferometer Gravitational wave Observatory), is being built by a consortium of universities in the USA, the principal partners of which are the California Institute of Technology in Pasadena, California, and the Massachusetts Institute of Technology. This instrument has interferometric arms that are 4 kilometres long. Two instruments will be built, one in the state of Washington and the other 3000 kilometres away in the state of Louisiana. The two instruments are required to eliminate signals due to local disturbances like seismic movements. A French–Italian collaboration is building a smaller interferometer at Pisa, Italy – the VIRGO project. This interferometer is 3 kilometres long. The GEO-600 project is a German–British collaboration to build a 600-metre interferometer near Hanover, Germany. A Japanese project called TEMA is a 300-metre interferometer at the National Astronomical Observatory, Mitaka, Tokyo.

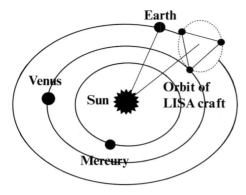

FIGURE 6.8 A schematic diagram of the three LISA spacecraft in an orbit 20 degrees behind the Earth. The three spacecraft are arranged at the corners of an equilateral triangle with sides 5 million kilometres long. The LISA configuration is effectively two giant interferometers.

An ambitious project proposed jointly by the European Space Agency and NASA is to place a large interferometer in space. The LISA (Laser Interferometer Space Antenna) proposal consists of three spacecraft orbiting the Sun about 20 degrees behind the Earth (Figure 6.8). The spacecraft are arranged at the corners of an equilateral triangle with sides that are 5 million kilometres long. The sides act as the arms of two giant interferometers. Each spacecraft is in an Earth-like orbit with a period of one year. The plane of the orbit of each spacecraft is slightly elliptical and slightly tilted with respect to the orbit of the other two spacecraft and with respect to the plane of the Earth's orbit (the ecliptic). By carefully choosing the tilts of the orbits, the three spacecraft maintain a triangular formation even though each is orbiting the Sun independently. Gravitational waves passing through the solar system will generate small changes in the distance between the spacecraft, which LISA instrumentation will measure.

LISA is up against all the challenges of a ground-based interferometer plus many more. For LISA to operate effectively all possible disturbances, apart from those produced by the gravitational waves, must be eliminated or calibrated out. The known source of disturbance in space is the Sun: both sunlight and solar wind will exert pressure on the three spacecraft. This pressure, if not corrected, will force the spacecraft to move distances much larger than those produced by gravitational waves. To reduce this effect, each spacecraft contains within it a 'proof' mass that is shielded from the Sun but is in no way connected to the surrounding spacecraft. The laser interferometer measures the distance between these shielded proof masses. The position of each spacecraft is measured relative to the proof mass within it. If the spacecraft moves with respect to the proof mass, small thrusters are fired to correct this movement.

Each spacecraft consists of two identical optical assembles, and lasers. Each optical assembly contains a 30-centimetre diameter telescope for transmission and reception of laser signals. An optical assembly also holds an inertial sensor containing the proof mass, laser detection and beam shaping optics, and detectors for measuring the position of the spacecraft with respect to the proof mass. In the 'basic' mode of operation, one spacecraft serves as the source of the laser beam and the detector of the interferometer (it also acts like the beam splitter in a ground-based interferometer). One of the optical assemblies in this primary spacecraft is chosen as the reference instrument. The laser of the reference optical assembly is locked to the cavity on its optical bench. The laser of the other optical assembly on the same spacecraft is also locked to the reference cavity by means of the optical fibre connecting the two optical assemblies. Each optical assembly of the primary spacecraft is pointed at one of the other two spacecraft of the triangle. The other two spacecraft serve as the test masses of the interferometer. In a ground-based interferometer the laser beam is reflected back to the detector off the two test masses. This is not possible for LISA as the arms are exceptionally long. To compare the length between the two proof masses in the spacecraft most of the light from the primary craft is transmitted to the secondary spacecraft, light that is retained serves as the local reference. The secondary spacecraft receives the beam from the primary, reflects it off the proof mass in its optical assembly and locks its laser to this beam. This laser then sends a beam back to the primary spacecraft. The primary spacecraft reflects the incoming beam off its proof mass and compares it with the reference. By thus bouncing the laser beams off the proof mass the measurement of separation of the proof mass is, to first order, independent of the motion of the surrounding spacecraft.

The instrumentation of the three spacecraft is identical and enables each spacecraft to act as either the primary or the secondary craft. This configuration preserves the basic interferometer even if on one spacecraft one optical assembly with its associated instrumentation fails. Disaster would ensue only if both optical assemblies were to fail on the same spacecraft. With all six optical assemblies fully functional, two independent interferometers can be formed, and these provide additional scientific information especially about the polarization of the observed gravitational waves.

The incredibly long arms of LISA will not only increase the sensitivity of the instrument considerably above that possible with ground-based interferometers but will also make the interferometer sensitive to low frequencies. These low frequencies, believed to be emitted by super-massive black holes, are drowned out on Earth by seismic activity. LISA is expected to achieve a sensitivity of 1 part in 10^{23}, especially at low frequencies.

Gravitational waves have been neither produced nor detected in the laboratory. Production of detectable gravitational radiation in the laboratory is impossible; movement of very large masses would be necessary. The only realistic sources of gravitational radiation are astronomical objects. Successful detection will not necessarily be a confirmation of the theory of general relativity because most alternative theories of gravitation also predict gravitational radiation. However, the properties of this radiation will be a test of general relativity and provide constraints on the alternative theories. Also, the detection of gravitational radiation will open up an entirely new chapter in observational astronomy comparable to the chapters opened by radio and X-ray astronomy. This is because although the waves are difficult (at present impossible) to detect, the amount of power in these waves is immense. This huge power means that these waves carry the imprint of the most energetic events in the universe, and they can propagate through space without dissipation, making them valuable carriers of information of events that are not observable by any other means. Also, because gravitational waves are emitted by the bulk motion of the source and not by individual atoms, they carry a completely different kind of information about their source. For example, the polarisation of gravitational waves from a binary stellar system reveals the inclination of the orbit to the line of sight, a crucial unknown in the modelling of such systems.

The most likely sources of gravitational waves are:

- Supernova collapse. Gravitational waves interact with matter very weakly and they are neither attenuated nor scattered (they can be gravitationally lensed) and can therefore reveal information about the hidden interior of a supernova explosion.
- Compact binaries and their coalescence.
- Black holes, their formation and coalescence. Gravitational waves provide the only way to observe black holes directly. It is the only radiation emitted with observable strength by these objects. At present all information about black holes is indirect, coming from their effects on the surrounding gas.
- The ambient background produced at the formation of the universe. The cosmic microwave background carries the map of the Universe as it was about 300 000 years after the big bang (see Chapter 7, Hubble & Eddington), and the studies of nucleosynthesis reveal conditions in the Universe a few minutes after the big bang. Gravitational waves on the other hand were produced soon after Planck time (see Chapter 7, Hubble & Eddington) and have travelled almost unimpeded through the Universe since then. Observation of this background would provide information about the crucial early moments of the big bang.

The sources of gravitational waves listed above are based on our current knowledge of astronomical phenomenon, but it should be recalled that unexpected discoveries have been made after opening every new window in the electromagnetic spectrum. We should not be surprised if there are objects and events out there which are only 'visible' via gravitational waves and these are waiting to be discovered in this millennium.

7 Hubble & Eddington

Gravity is the only truly universal force. It moulded the universe and it almost certainly will overwhelmingly determine the end of the universe. Today we also know (certainly believe) that gravitation is primarily responsible for the formation of the large structures we see around us: the Earth, the solar system, the stars, and the galaxies. Gravity has fashioned the beautiful and awe-inspiring sights in the sky, which have inspired both philosophers and mystics. Without gravity the sky would have been a very boring sight. Even more exotic objects, only visible at radio, X-ray or other energies, are present in the sky and these have also been fashioned by gravity. After the formation of the solar system, gravity has played a benign role in the evolution of life in the solar system. The strong gravity of Jupiter has shielded Earth from destructive impacts by comets and asteroids, and it is reasonable to say that life on Earth would not have survived without this 'gravity shield'. It is for this reason that there is such excitement at the discovery of large planets around other stars. Life, as we know it, may not (almost certainly does not) exist on these large planets, but without such a large planet and its gravity shield, life certainly would not survive and flourish on an inner planet if there is one. In this chapter the central role of gravity in shaping the universe and even the climate on Earth is described.

FIGURES OF PLANETS
The sizes and the figures of the planets have been subjects of inquiry and discussion since antiquity. The Greeks accepted the Platonic

belief in perfect shapes and believed that the planets, and the Earth in particular, were perfect spheres. By the Middle Ages this Platonic-Aristotelian view of a spherical Earth was lost and a flat Earth was in vogue around the time Columbus set sail in 1492 to 'discover' the Americas. In chapters two and three of *De Revolutionibus* Copernicus gave detailed arguments for a spherical Earth, and by the seventeenth century a spherical Earth was 'established'. In 1669/71 the French Académie des Sciences organised the measurement of the metric length of a degree on Earth's surface. Jean Picard (1620–1680) made this measurement by observing the position of a star from two locations. Picard showed that a degree was 69.1 miles (111.2 kilometres) in 'length'. This value was very different from the 60 miles that had been generally accepted[1]. Picard's measurement was published in 1671 in an obscure journal and was missed by Newton.

Soon after Picard's determination of the length of a degree, the Académie organised a geodetic expedition to Cayenne in French Guiana (South America). Cayenne is 5 degrees north of the equator and it was found that to keep time a pendulum set for Paris (49 degrees north) had to be shortened. We now know that this is because the Earth bulges at the equator: an observer there is further from the centre of the Earth and therefore the gravitational attraction decreases. Because of this weaker attraction a pendulum of a given length swings more slowly at the equator. The results of the Cayenne expedition were published in 1684. In the *Principia* Newton used these data to determine the figure of the Earth.

In Book III of the *Principia* Newton discusses the figure of the planets and of the Earth. He starts from the premise (given in Book I) that equal gravitational attraction by material on all sides will result in a spherical shape for the planets. But he noticed that:

> the diameter of Jupiter is found shorter between pole and pole than from east to west

Newton then conjectured that the same would be true of the Earth. This is because in 1685 the figure of the Earth had not been observationally established; it was a further 50 years before the true shape was determined. Newton showed theoretically that the oblate shape (flattened at the poles) of a planet was the result of the combined effect of gravitation attraction and the centrifugal force due to the rotation about the axis of a planet. Between 1684 and 1714 a series of pendulum measurements were made in France by Giovanni Domenico Cassini and his son Jacques, to determine the Earth's figure. The results of these measurements suggested that the Earth was a prolate spheroid (extended at the poles), in disagreement with the theoretical work of Newton and Huygens. Thus the figure of the Earth became a major subject of scientific research in the early eighteenth century and a number of expeditions were organised to make pendulum measurements at various locations. Of these, the two most important were those made in 1735 by de la Condamine (1701–1774) in South America, in the neighbourhood of the equator, and those by de Maupertuis in 1738 in northern Sweden, within the Arctic circle. The results of these expeditions established that the Earth was indeed an oblate spheroid as computed by Newton.

TIDES

The twice-daily high and low tides in the oceans have also been known since antiquity. Kepler was first to attribute a nonterrestrial cause to the tides; he believed that the Moon was responsible for tides. Galileo dismissed this (in a characteristic Galilean manner) as absurd. He believed, wrongly as it turns out, that tides were caused by the Earth's two motions (one around its own axis, the other around the Sun), like water sloshing about in a moving container. Newton provided an explanation of the most significant characteristic of the tides within the context of his theory of gravitation. Newton's theory of tides is qualitatively correct but is inadequate to predict the time or the height of the tides at any particular place. This is because the theory of tides is not a problem in statics, as presented by Newton, but one of dynamics. A dynamical theory of tides has to include the rotation of the Earth and this is not trivial. Laplace developed the correct theory of tides a century after Newton.

The oceanic tides are a consequence of centrifugal force due to the Earth's rotation together with gravitation in the Earth–Moon and Earth–Sun systems. In the Earth–Moon system both the Earth and Moon orbit around their common centre of mass. This common centre is about 4700 kilometres from the centre of the Earth (for comparison, the equatorial radius of the Earth is 6378 kilometres). The Earth rotates about this point with a period of one month. The Moon's gravitational attraction and the centrifugal force due to the rotation of the Earth are exactly balanced at the centre of the Earth. At the sub-lunar point on the surface of the Earth, the gravitational attraction of the Moon is stronger than the centrifugal force of the Earth's rotation, and the ocean's waters, which are free to move, build up in a bulge in response to this unbalanced force. On the surface of the Earth opposite the sub-lunar point the centrifugal force is stronger than Moon's gravitational attraction and another small bulge of water develops. The ocean's waters are correspondingly depleted at points on Earth's surface perpendicular to the Earth–Moon line. Each day the Earth rotates beneath these bulges and troughs, which remain stationary relative to the Earth–Moon line. The result is two high tides and two low tides every day when a location on the Earth is carried into the bulges or the troughs by the rotation. At new and full Moon the Moon and the Sun act together on the ocean's water to produce the large spring tides. The neap tides occur at first and last quarter. Like the ocean's waters the solid body of the Earth also experiences twice-daily tides with a maximum amplitude of 30 centimetres.

In the nineteenth century the British astronomer-geophysicist, Sir George Howard Darwin (1845–1912), second son of Charles Darwin, showed that because of the viscosity of water the high point of a tide is always in advance of the position it would reach if water were a perfect fluid. He also pointed out that the dissipation of tidal energy slows down the Earth's rotation and the Moon's orbit gradually expands. The length of the day increases by 0.0016 seconds per century and the Moon recedes by 2.6 centimetres per year.

In addition to the Earth–Moon system, numerous other consequences of tidal dissipation can be observed in the solar system and elsewhere in the Milky Way. The spin of Mercury has been slowed down by tidal interaction with the Sun and the angular velocity of Mercury is now exactly 1.5 times its orbital mean motion. Owing to tidal interaction all major planetary satellites but one are observed to be rotating synchronously with their orbital motion. The exception is Saturn's satellite Hyperion. Tidal friction has retarded Hyperion's initial spin rate to a value near that of synchronous rotation, but the combination of the satellite's unusually asymmetric shape and its high orbital eccentricity leads to gravitational torque that makes the synchronous rotation unstable. As a result Hyperion now tumbles chaotically with large changes in the direction and magnitude of its spin on time scales comparable to its orbital period of 21 days.

Differential tidal expansion of orbits has also maintained several orbital resonances. The orbital resonance among Jupiter's satellites Io, Europa and Ganymede, whose orbital periods are nearly in the ratio 1:2:4, maintain Io's orbital eccentricity at the value of 0.0041. This low value of eccentricity causes large variations in the magnitude and direction of Io's enormous tidal bulge and the resulting dissipation of tidal energy melts a significant fraction of the satellite. As a result Io is the most volcanically active body in the solar system. The orbital eccentricity would normally have been damped to zero by this large dissipation, but the orbital resonance with Europa and Ganymede prevents this from happening.

The tides raised by the Sun and other planets are negligible at the distance of Pluto and its satellite Charon and it is possible that further tidal evolution of this planet system has ceased. In this state the orbits are circular, with Pluto and Charon rotating synchronously with spin axes perpendicular to the orbital plane.

ICE AGES

Over the last million years the Earth has experienced huge natural climatic changes as it has switched into and out of ice ages. Geological evidence shows that in this rhythm, an ice age several thousand years long is succeeded by a warmer interval, ten to twenty thousand years long, called an interglacial period. In the nineteenth century the British scientist James Croll proposed an astronomical theory of these ice ages. But this work was not generally accepted till it was rediscovered in 1930s by the Yugoslav astronomer, Milutin Milankovitch. Milankovitch was a Serb; caught in the wrong country at the outbreak of the First World War, he was interned by the Austraian authorities. Luckily his plight came to the attention of a friendly Hungarian professor who had him paroled and moved to Budapest in Hungary. At the library of the Hungarian Academy of Science Milankovitch continued his meticulous calculations of the Earth's motion and identified three principal astronomical cycles which would affect the warming and cooling of the Earth. Today the astronomical theory of the cyclical ice ages is called the 'Milankovitch model'.

The Earth rotates on its axis in 24 hours and goes round the Sun in about 365 days. But the axis of Earth is not at right angles to the ecliptic plane, that is the plane of the Earth's orbit. The axis is tilted at about 23.5 degrees to the ecliptic (Figure 7.1) and at present the axis 'points' to the Pole Star. This tilt of the Earth's axis is rather fortunate because it adds variety to the annual weather pattern; without this tilt there would be no seasonal change in the weather and every day of the year would have 12 hours of daylight and 12 hours of night. When the Sun is on the same side of the Earth as the Pole Star, there is summer in the northern hemisphere and 24 hours of sunlight at the north pole. Six months later the Earth's axis is tilted away from the Sun and it is winter in the northern hemisphere and summer in the southern hemisphere and the south pole has sunlight for 24 hours. Spring and autumn occur when the Earth is at intermediate positions in its orbit round the Sun.

The spinning Earth is like a gyroscope and on short time scales the tilt of its axis is fixed. However, because of the (small) gravitational forces due to the Sun, Moon and other planets acting on Earth, the axis precesses around the line perpendicular to the ecliptic (Figure 7.1). This is the 'precession of the equinox', which was discovered by ancient Babylonians and later by Hipparchus (see Chapter 1, Aristotle). The period of this precession is about 26 000 years. Because of this precession the pattern of seasons changes slowly over a few thousand years.

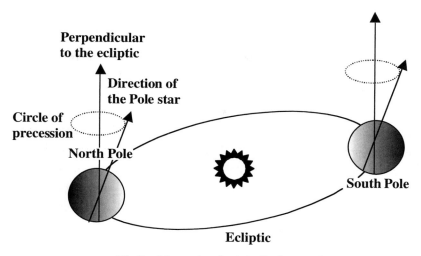

FIGURE 7.1 The Earth's rotational axis is tilted at 23.5 degrees to the line perpendicular to the ecliptic. The seasonal climatic changes are due to this tilt. The change in the tilt and in the shape of Earth's orbit account for long-term changes in the climate and for the ice age cycles.

However, the tilt angle of the axis does not stay fixed. Small changes in the gravitational perturbations cause the axis to nod towards and away from the ecliptic from 21.8 degrees to 24.4 degrees. The period of this cycle is about 41 000 years. The current tilt of 23.5 degrees is about halfway between these two extremes. At present the tilt is decreasing and the difference between summers and winters is less today then it was a few thousand years ago.

The third component of the Milankovitch model is the slight change in the Earth's orbit around the Sun. This change is also due to slight changes in the gravitational forces in the solar system and results in the Earth's orbit stretching from circular to elliptical and back with a period of about 100 000 years. Over this cycle, the distance between the Earth and the Sun varies by as much as 18.26 million kilometres. These three periodic changes affect the amount and the angle of the solar radiation received at Earth in different seasons. Thus over long periods the amount of heat in the Earth's atmosphere is redistributed although the total amount of heat in the atmosphere does not change significantly.

According to the Milankovitch model it is this redistribution of heat that is responsible for triggering the periodic ice ages on Earth. The Milankovitch model was considered extremely implausible when it was first proposed, as the change in the seasonal heating is small. However, climatic temperatures of various past epochs determined from cores drilled from the deep-sea floor and ice cores from the Antarctic have supported the link between ice ages and the astronomical cycles. The climatic changes over the past 800000 thousand years have now been determined and the fundamental role of the three astronomical cycles in the observed changes has been established. Interestingly the dominant cycle is the 100000-year eccentricity cycle, which was expected to be the weakest of the three cycles. It is now known that various biological and chemical processes in Earth's atmosphere and oceans are amplified by the gravity-induced changes in the distribution of heat and these trigger the ice ages and the interglacial periods.

THE UNIVERSE — AN EXPANDING FIRE-BALL

From time immemorial the universe has aroused wonder and curiosity in all cultures. From the seventeenth century onwards scientists increasingly began to interpret the universe in terms of the mechanical world-view of Newton. Newton himself had shown that the application of his theory of gravitation led to an unstable universe. In the late nineteenth century attempts were made to remove the instabilities inherent in Newton's universe by modifying the law of Newtonian gravity. This was done by adding a very weak *repulsive* component in Newton's law. This kind of universe received strong support from observations of stellar density, which appeared to decrease with distance from the centre of the Milky Way. This universe, like Newton's universe, was a static universe.

Einstein planted the seed of modern cosmology on 8 February 1917 when he presented a paper in which he explored the consequences of applying his general theory of relativity to the entire universe. In this application of general theory, Einstein made two assumptions: (1) the universe is homogeneous (all parts are of the same kind) and isotropic (physical properties are same in all directions) and (2) the mean density and curvature are constant. The first assumption is now enshrined in the Cosmological Principle; this hypothesis is required in order to avoid a privileged observer. The second assumption was supported by the empirical data then available. These data suggested that the universe was spatially finite and static, and it contained a finite amount of matter. To his chagrin Einstein found that the solution of his cosmological equation was not consistent with the second assumption – the solution suggested an expanding (or a contracting) universe. This is a consequence of the positive curvature of space-time implied by the attractive nature of gravity (see Chapter 5, Einstein). This flew in the face of 'common knowledge'; that is, that the universe was static, unchanging and eternal, something that 'had been known for centuries'! To his great shame, Einstein, the revolutionary, the first person to successfully challenge the Aristotelian concepts of absolute space and time, and the person who swept aside the Newtonian concept of gravitation, lost his nerve. He introduced a 'fudge factor' – the cosmological constant – to prevent the expansion of the universe. Einstein admitted in his 1917 paper that this was not justified by the theory of gravitation but was necessary to maintain a quasi-static distribution of matter – as observed. The cosmological constant is a cosmic repulsive force that increases with distance. The universe, in Einstein's 1917 model, is a balance between the attractive (Newtonian) gravitational force and the repulsive cosmological force denoted by the cosmological constant. The value of the constant is not known, but the solution of Einstein's field equation *without* the cosmological constant is known to be in excellent agreement with observations within the solar system, so any such constant must be extremely small. This 'modification' of general relativity was conceptually similar to the modification that had been introduced in the Newtonian theory to 'stabilise the universe'.

Einstein's universe was spatially closed and finite but temporally infinite. This model is referred to as Einstein's 'cylinder' universe: a model universe in which the axis of the cylinder represents the time coordinate and the radius represents the space coordinate (two other spatial dimensions are suppressed, for visual clarity). Einstein found the cosmological constant aesthetically unsatisfying but he could see no alternative to it. In 1919 Einstein admitted that the introduction of the cosmological constant was 'detrimental to the formal beauty of the theory'. Twelve years later new astronomical observations suggested that the universe was not static after all but was expanding and Einstein was to admit that the introduction of the cosmological constant had been a mistake. Whatever the rights and wrongs of the cosmological constant, the idea of treating the universe by a relativistic field equation was a revolution in the age-old concept of the universe. The universe was no longer confined by the observable boundaries (determined by the limits of technology) but was open to mathematical interpretation in both time and space. This new opportunity to study the universe was not immediately seized by astronomers and physicists; cosmology remained an esoteric discipline practised by a few specialists.

Although Einstein had abandoned the cosmological constant in 1931, it has not gone away. Observations published in 1998 (described later) suggest that supernovae at high redshift are fainter than expected (from the application of the standard theory of Einstein). This implies that they are further away than expected or the space has stretched out more than expected. This is possible if the cosmological constant is larger than zero resulting in slight opposition to gravity.

Einstein presented his cosmological model in Germany during the First World War. The model was 'brought' to Britain and presented to the Royal Astronomical Society by the eminent Dutch astronomer Willem de Sitter (the Netherlands were neutral during the First World War and Dutch scientists were able to keep in touch with scientists in Germany and other European countries). De Sitter was at home with both theoretical and observational astronomy and he went further than just presenting Einstein's work. He showed that the matter-filled universe was not the only solution to Einstein's cosmological equation but a static solution was also possible for a universe empty of matter. Moreover, he showed that in his universe radiation emitted by a distant source would appear redshifted to an Earth-based observer and this redshift would increase with distance. He described this as a 'spurious' positive radial velocity because it was not a real velocity caused by expansion of space; after all, his was a static model. The redshift was a consequence of the space-time that described de Sitter's universe. Einstein was not happy with the de Sitter model. He accepted the mathematics of it but he considered a universe empty of matter to be unrealistic. De Sitter, on the other hand, maintained that even in a matter-filled universe the density of matter would be so low that an empty universe was a good approximation. In the early 1920s these were the only two available models of the universe.

Unknown to the Western scientific community the Soviet physicist Alexander Alexandrovich Friedmann (1888–1927) had analysed non-static solutions of Einstein's cosmological equation. Friedmann was born in St Petersburg (then Petrograd, later Leningrad and now again St Petersburg) on 17 June 1888. In the local gymnasium (primary and secondary schools) Friedmann showed great interest and talent for mathematics and in 1906 he entered St Petersburg University to study pure and applied mathematics. At the University he came under the influence of Paul Ehrenfest, Einstein's friend and colleague, who taught in St Petersburg until he went to Leiden in the Netherlands in 1912. During the First World War, Friedmann served on the Austrian front. He survived the War and the following Soviet Revolution to become, in 1925, the director of the Main Geophysical Observatory in St Petersburg. Unfortunately he died in 1927, at the young age of 39. Friedmann turned to the study of relativity around 1920. In 1922 he offered a complete analysis of the solutions of Einstein's cosmological field equations. Friedmann recast Einstein's equations to resemble the equation of motion of a point particle on the surface of a sphere. He took his analysis beyond the solutions of Einstein and de Sitter by including nonstatic solutions. He found a class of solutions that, depending on the value of the cosmological constant, included Einstein's 'cylinder' universe as well as a homogeneously expanding universe. Friedmann thus introduced into cosmology two fundamental (and revolutionary) concepts:

the creation of the universe and the age of the universe. The age of the universe was defined as the time that has elapsed from the moment when space was concentrated at a point (creation) to the present state. For the first time the idea of an expanding universe originating in a singularity was introduced in cosmology – the notion of a big-bang universe was born. Friedmann was first and foremost a mathematician, more interested in the mathematical structure of the solutions he had obtained, and he made no attempt, in 1922 or later, to connect his findings with astronomical observations.

Friedmann's paper was published in the world's leading journals of physics, was studied by several leading physicists and astronomers, and ignored. Einstein also read his paper and failed to appreciate or properly understand the message of the paper. Einstein first thought that there was an error in the paper and he claimed that it proved the static nature of relativistic world models. Friedmann was puzzled by Einstein's comments. He rechecked his calculations and wrote to Einstein explaining that there was no error. Einstein realised his own mistake, publicly retracted his objection and admitted that there were indeed time-varying solutions to the field equations. But this was acceptance of the mathematics of the nonstatic solutions; Einstein appears to have failed to understand their physical significance. In a paper completed in 1929 the American physicist Howard Percy Robertson referred to Friedmann's work in a footnote. Robertson had studied Friedmann but he had also failed to recognize the significance of the evolutionary solutions. This was to change a year later when, in the light of new observations, Friedmann's paper was seen as a brilliant prediction of the expanding universe.

In 1929 the American astronomer Edwin Powell Hubble (1889–1953) published his now famous diagram of velocities of galaxies plotted against their distance (Figure 7.2). Edwin Hubble was born on 20 November 1889 in Marshfield, Missouri, USA. the third of seven children who survived. In his school years he demonstrated both academic and athletic abilities and at the age of 16 entered the University of Chicago. He graduated four years later with high grades in science subjects and was awarded a Rhodes scholarship to study in Britain. In 1910 he arrived at Queen's College of Oxford University to read law. Three years later he returned to the USA to practice law at Louisville in Kentucky (where his family had moved). But Hubble's heart was not in law and in 1914 he returned to the Yerkes Observatory of University of Chicago to start graduate studies in astronomy. At the first meeting of the American Astronomical Society that he attended, he was fortunate to be present at the talk given by the pioneering American astronomer Vesto M. Slipher. Slipher had obtained the first well exposed and calibrated spectrum of a spiral nebula. This spectrum was characteristic of a collection of stars. But more importantly, Slipher showed that the nebular absorption lines were redshifted (i.e. shifted to longer wavelengths). This Doppler shift indicated that the nebula was moving away at a considerably higher speed than the typical speeds of stars in the Milky Way. Slipher's data strengthened the argument in favour of the extragalactic nature of spiral nebulae. By 1925 Slipher had obtained spectra of 45 nebulae, 41 of which were redshifted.

Inspired by Slipher's talk, Hubble started a programme to photograph the nebulae with the Yerkes Observatory's 24-inch refracting telescope. He used these data to classify the nebulae and showed that most of the nebulae were elliptical and not spiral, and many of them appeared to be clustered together and were more likely to be seen away from the plane of the Milky Way. This work earned Hubble his Ph.D. In 1916 he accepted a position at the new Mount Wilson Observatory in Pasadena, California. However, his move to Mount Wilson was interrupted by US entry into the First World War in April 1917. Hubble immediately joined the army; he had obtained his doctorate just three days previously. After basic army training in the US he was shipped to France, but fortunately the war ended before he was sent into combat. He was discharged from the army in 1919 and immediately went to Mount Wilson. He arrived at an opportune time as the observatory had a 60-inch and the new 100-inch (then the largest in the world) reflecting telescope. At Mount Wilson he continued his research on the nature of spiral nebulae. In 1923 he identified a number of variable stars in the nebula M31, the Andromeda galaxy. The Andromeda galaxy is the only galaxy visible to the naked eye (from the northern hemisphere) and had been noted as a fuzzy patch of light by the tenth-century Persian astronomer 'Abd ar-Rahman al-Sufin (see Chapter 1, Aristotle). Hubble determined that the stars in question were Cepheid variables. Cepheids are a class of stars whose brightness varies regularly, a phenomenon known since 1786. In the early 1900s Henrietta Leavitt, working at the Harvard Observatory, discovered similar variable stars in the Magellanic Clouds, our nearest galactic neighbour. In 1912 Leavitt determined that there was a linear relationship between the period and the apparent brightness (magnitude) of this class of stars in the Large Magellanic Cloud. This was an important discovery because it could be used to determine the distance to these stars. Between 1913 and 1918 the Danish astronomer Ejnar Hertzsprung and the American astronomer Harlow Shapley determined the relation between the period and luminosity of Cepheids. This allowed a relation to be established between the period and the distance to the Cepheids. Hubble used the period–luminosity or period–distance relation to determine the distance to M31. By 1924 he had determined the distance to another nebula, M33. These results were presented by the astrophysicist Henry Norris Russell at the 1925 joint meeting of the American Astronomical Society and the American Association for Advancement of Science. Hubble had shown that both these nebulae are about 930000 light-years away, well beyond the bounds of the Milky Way. Hubble had confirmed the conjecture of the eighteenth-century philosopher Immanuel Kant that some nebulae were 'island universes'. These extragalactic nebulae are now known as galaxies.

By 1929 Hubble had reasonably reliable distances to 46 galaxies whose redshifts had been measured by Slipher or Milton Humason. Humason was an exceptionally unusual astronomer; he had no formal education beyond the primary school, and had worked as a mule driver and a foreman on a California ranch. He became a janitor at the newly built Mount Wilson Observatory and by 1921 had advanced to become a spectroscopist at the observatory. In 1929 he measured the redshift of the

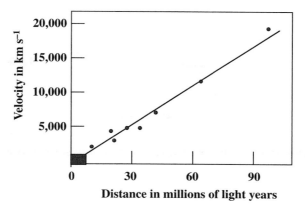

FIGURE 7.2 The distance–velocity diagram for galaxies or the Hubble diagram. By 1931 Hubble and Humason had extended this diagram out to about 100 million light-years or about 30 megaparsec. The 1929 data of Hubble would have covered just the region shown by the shaded patch at the bottom left of the figure.

galaxy NGC (New Galactic Catalogue) 7619 and reported a radial velocity of 3779 kilometres per second. This was twice as large as the previous record for a galaxy's redshift. Hubble used these data to establish a relation between radial velocity and distance and in spite of the large scatter in his data, he claimed that there was a linear relation between recession velocity and distance to the galaxies (Figure 7.2). Hubble presented his work as an empirical investigation, claiming that observational data could be introduced in the discussion of cosmological models. Hubble was not a theoretician but he was aware of the cosmological theories and the distance–redshift relation of the de Sitter model. By 1931 Hubble and Humason had obtained enough data to show that the linear relation between recession velocity and distance was valid out to 100 million light-years or about 30 million parsecs (megaparsec) as shown in the figure[2]. The Hubble law ($v = H_0 r$) is now seen as a reality. The Hubble constant H_0 is expressed as kilometres per seconds per megaparsec (kms^{-1} Mpc^{-1}) and has dimensions of (time)$^{-1}$. The 1931 data of Hubble and Humason suggested a value of 558 kms^{-1} Mpc^{-1} for this constant, that is an age of the universe of about 1.8 billion years (the US (or French) billion, that is 10^9, is used in this book). It was realised that this was an 'interesting' number and should be compared with the 'cosmic age'. However, this was an embarrassment. Ernest Rutherford, the New Zealand physicist working at Cambridge University, had shown from measurements of the abundance and half-life of uranium isotopes, that the lower limit of the age of Earth was about 3.4 billion years.

Determination of the value of the Hubble constant has been a major goal of observational astronomy since the 1930s and vast amount of telescope time has been devoted towards this end. One of the primary scientific missions of the multi-billion dollar Hubble Space Telescope (HST) was to determine this constant. The measurement is fraught with difficulties as it depends on a complex inter-related chain of measurements to estimate the distance to remote galaxies. The principal link in this chain is the period–luminosity relation of Cepheid variable stars. In 1994 and 1995 research groups in the USA, Canada, Australia and the UK used the HST to observe Cepheids in the Virgo cluster of galaxies and in the Leo I local group of galaxies and obtained values of the Hubble constant between 87 and 69 kms^{-1} Mpc^{-1}. These values suggest a cosmic age of about 12 billion years. This age is comfortably higher than the age of the Earth. It is also comparable to the age of the stars in the Milky Way. The oldest stars in the disc of the Milky Way, estimated from the cooling rate of white dwarfs, are about 9 billion years old. The stars in the halo of the Milky Way are about 15–16 billion years old, according to the age derived from the rate of consumption of nuclear fuel in the core of these stars.

Friedmann made no attempt to link his solutions of Einstein's field equations with astronomy or physics; this was left to the Belgian physicist Georges Édouard Lemaître (1894–1966). Lemaître's early education, at the Catholic University of Louvain, was in engineering, not because he was interested in the subject but because he felt it would help him to support his family. However, the First World War interrupted his studies and he spent over five years, from August 1914, in the Belgian army. He was engaged in heavy house-to-house fighting and witnessed the first poison gas attack in the history of warfare. But in quieter periods he relaxed by reading books on physics. At the end of the war, in 1919, Lemaître returned to the University of Louvain to study physics and mathematics and later also theology. He was ordained a priest in 1923 and for the rest of his life he retained his two careers, one clerical and the other scientific. Lemaître was elected to the Pontifical Academy of Science in 1936 and was the president of the Academy from 1960 until his death in 1966.

Lemaître was attracted by complicated mathematical problems and was fascinated by the logical beauty, simplicity and unity of general relativity. He had an opportunity to cultivate this interest further in 1923 when he spent a year at Cambridge University, as a student of Eddington – an authority on relativity. Eddington was impressed by the great mathematical ability of the Belgian student. Lemaître spent the following year in the United States, where he worked with Shapley at the Harvard College Observatory and obtained a doctorate from MIT. Lemaître was quick to recognise the cosmological significance of the astronomical discoveries being made by Slipher and Hubble and he visited both these observational astronomers when he was in the United States. Back in Louvain, Lemaître made a systematic study of the relativistic world models and rediscovered most of the work done independently by Friedmann. There was, however, a significant difference. Lemaître solutions included a radiation pressure; he considered matter pressure to be negligible. This was the first time thermodynamics had been introduced into relativistic evolutionary cosmology. Unlike Friedmann, Lemaître was interested in developing a physically realistic cosmology and wanted to present a definite evolutionary model of a real universe and identify the cause for expansion. Lemaître suggested, tentatively, that the expansion was due to radiation pressure in the universe, but he was unable to develop this further. Lemaître also explicitly described his model as an expanding one and provided a natural explanation for the recession velocities of galaxies which Hubble and others had determined.

Lemaître's model of the universe was not a big-bang model; it did not have a definite age. The model evolved from an Einstein universe, which was considered an equilibrium state and which was somehow perturbed. The model then asymptotically approached the de Sitter universe. The big-bang universe was implicit in Lemaître's equations but he did not develop this in the late 1920s. It is possible he found these models unphysical. Einstein was as unenthusiastic about Lemaître's evolving models as he had been about Friedmann's work. He accepted the mathematics of these models but he regarded them to be physically irrelevant. Sadly Lemaître seems to have gone out of his way not to discuss his 1927 models with the leading relativists of the time. He also tended to publish his work in relatively obscure journals. By 1930 both Eddington and de Sitter had 'converted' to a nonstatic universe and Lemaître sent a copy of his 1927 paper to Eddington. Eddington realised to his embarrassment that three years earlier he had ignored (or failed to understand) a significant piece of work. But this oversight was quickly remedied: in a letter to the British science journal *Nature* Eddington drew attention to Lemaître's brilliant work. Eddington and de Sitter's enthusiastic endorsement including an English translation of the 1927 paper brought Lemaître's theory to the attention of the international science community. Almost overnight Lemaître became a scientific celebrity and the expanding universe became the 'standard model' of the universe. In this theory Lemaître proposed that the expansion of the universe could be traced back to an exceedingly dense state which he called the primeval 'super-atom'. Even Einstein, who only a couple of years earlier had dismissed Lemaître's theory, now accepted it enthusiastically and 'blessed' it publicly in a talk given in 1931. Einstein's 1931 model was of the big-bang type but he found the singularity (at the moment of formation) disturbing. He hoped that this singularity could be explained away as a mathematical artefact by assuming an inhomogeneous distribution of matter at small radii.

In the 1950s the Russian-American physicist Georgii Antonovich Gamow (1904–1968) decided to add nuclear physics to the Lemaître–Einstein evolutionary model of the universe. Gamow was born in Odessa in southern Ukraine. As a young boy he decided to look for proof of transubstantiation by examining under his small microscope the wine and bread served during Communion. He was disappointed to find no proof. His early education was during the tumultuous political and military upheaval of the Bolshevik Revolution in the Russian Empire but this appears not to have affected or interested him in the slightest. He went to university in Petrograd where he came under the influence of the giants of Soviet physics. In 1923–1924 he attended lectures in general relativity and this fired his interest in relativistic physics. Gamow graduated from the University of Leningrad in 1928 and went on a fellowship to Göttingen, Cambridge and Copenhagen. It was during his visit to Göttingen that he discovered the mechanism for expulsion of alpha particles from a nucleus, a discovery that propelled him onto centre stage in international physics. Gamow often travelled outside the former Soviet Union visiting Western Europe and working on nuclear theory, a field in which he was a pioneer and an expert. In 1932 the political climate in the Soviet Union deteriorated and Gamow decided to emigrate. He arrived in the United States in 1934 and took a position at the George Washington University.

Gamow was not interested in the mathematics of the space-time of the early universe, his interest was in the evolution of matter in this universe and formation of the elements. Gamow (and his collaborators Alpher and Herman) assumed that the universe started with a primordial super-dense state of nucleonic matter. This universe was dominated by radiation and Gamow believed that the remnant of this primordial radiation would still exist as a low-temperature microwave background radiation (more about this later). In the Gamow–Alpher–Herman theory, the elements observed today were formed in this universe as it expanded and cooled. In spite of early successes of the Gamow big-bang theory the theory failed to gain general recognition and work on it stopped till about the mid-1960s, when particle physicists realised that the high-energy reactions observed in particle accelerators might have occurred naturally in the early universe. The basic tenets of Gamow's big-bang theory are now accepted but it is recognised that only the light elements (hydrogen, deuterium, helium, lithium etc.) were formed in the early universe. The heavier elements are formed in stellar atmospheres. In particular the high abundance of helium (about 25% by mass of hydrogen) can only be formed in a high-temperature universe a few minutes after the big bang. This was discovered by the British physicists Fred Hoyle and his collaborator Roger Taylor. It is ironic that Hoyle, one of the originators of the 'steady-state' theory and a consistent critic and opponent of big-bang cosmology, should have discovered a crucial aspect for this model.

According to the current big bang[3] theory the universe was formed in a cataclysmic event of something and somehow at about 15 billion years ago. In this incredibly hot, dense event, space, time, matter and energy were all created. At present practically nothing is known about what happened in the first 10^{-44} seconds (the Planck time) of the universe. When Saint Augustine (354–430 AD) was asked what God was doing before he created heaven and the Earth, he replied, 'I keep away from the facetious reply. . .'. This is perhaps the answer a cosmologist might (or should!) give if asked about the history of the universe before Planck time. There is no point in appealing to physics, as physics did not exist at that time – certainly not physics as we know it. After the Planck time physics is on marginally secure grounds. At and before the Planck time, the four forces of nature (gravitation, electromagnetism, weak and strong) were (it is believed) united into one force whose nature is, at present, unknown. At about the Planck time the temperature of the universe dropped to about 10^{32} degrees and gravity separated out of this primordial single force. This would be the last time gravitons, the particles associated with the gravitational field (see Chapter 8, Planck), would interact with the surrounding space-time. After this time the gravitons would move through the expanding universe without interacting and carry information of the epoch immediately after Planck time. The history of the universe for next three minutes was determined by three phase transitions (which resulted in the remaining three forces of nature) and gravity resisting the expansion driven by the initial high temperature.

It is misleading to interpret the big bang as an explosion of matter

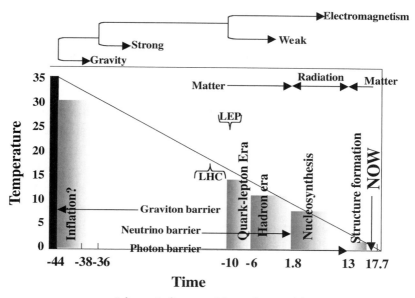

FIGURE 7.3 Schematic diagram of the evolution of the universe. The drop in temperature is plotted as a function of time in seconds; temperature and time are both given in logarithmic units (so 30 means 10^{30}). For detailed explanation of the diagram see text.

away from a particular point in space. The observed expansion was the expansion *of* the created space and not expansion *into* existing space. Time also began with the formation of the universe; that is, the physical universe came into being with time and not in time. The radial velocities of galaxies measured by Slipher, Humason and Hubble (and others) can now be interpreted as the motion of galaxies – the Hubble flow – with the expanding space. This is perhaps best visualised as the surface of a partially inflated balloon, with dots on it which represent the galaxies. As the balloon is inflated further the dots recede from each other. In the big bang universe space is dynamic and not just a passive background; as it expands the galaxies move with it, *not in it*.

The history of the universe from the Planck time onwards is shown schematically in Figure 7.3. When the universe was about 10^{-38} seconds old it is believed to have gone through an exponential expansion increasing in size by a factor of about 10^{50}. This is called the 'inflation' phase of the universe. A version of inflationary theory was first proposed in 1979 by Alexis A. Starobinsky of the Landau Institute of Theoretical Physics in Moscow. The theory was subsequently developed by Andre Linde and David Kirzhnits of the Lebedev Physics Institute in Moscow, Alan Guth of MIT and Paul Steinhardt of University of Pennsylvania. The theory explains the amazing uniformity and the lack of curvature of the observed universe. In this theory, towards the end of the inflation phase the energy of the space-time is converted to particles through Einstein's $E = mc^2$ and the universe passes into the first matter-dominated era. At about 10^{-36} seconds the temperature drops to about 10^{28} degrees and the strong nuclear force separates out of the combined two nuclear forces and the electromagnetic force. This separating of forces is known as 'symmetry breaking'. By 10^{-10} seconds the particles known as quarks, leptons and electrons begin to form from photons. This is called the quark–lepton era. Temperature is a measure of the average kinetic energy of the particles. It is therefore possible to relate the temperature at these early epochs to the energy of particles in accelerator experiments. The particle energies in the universe at 10^{-10} seconds have been reproduced in the Large Electron Positron collider near Geneva in Switzerland, and the particle interactions of the quark–lepton era have been studied in great detail. Earlier periods in this era will be probed with the Large Hadron Collider (at present the LHC is expected to be operational in 2005). At about 10^{-6} seconds the temperature drops to about 10^{12} degrees and the weak nuclear force and the electromagnetic force separate, so there is now the full complement of the four forces of nature. The temperature is low enough for quarks to begin clumping. Triplets of quarks form protons and neutrons (also antiprotons and antineutrons). Annihilation of matter and antimatter starts, eventually leaving a slight residue of matter by a process as yet unidentified. This is called the hadron era. As the temperature drops further, neutrons, protons and leptons are in equilibrium through weak interactions. The possible conditions during the hadron era have been investigated in some detail at various collider experiments. Between 1 to 3 minutes after the big bang the temperature drops to about 10^9 degrees and protons and neutrons can no longer be in equilibrium (through radioactive decay of the proton) and the neutron-to-proton ratio 'freezes' at 3 to 2. Cosmic neutrinos interact for the last time with the matter at this epoch. This era is dominated by radiation and nucleosynthesis begins. Hydrogen, deuterium and helium nuclei are synthesised and the helium abundance increases to about 25% (by mass) of hydrogen abundance. At about 300 000 years the temperature drops to about 3000 degrees and this is cool enough for neutral hydrogen and helium to form. The energy of radiation decreases (that is, the radiation is redshifted to lower energy owing to the expansion of the universe) and it is unable to ionise hydrogen; matter and radiation separate and go their independent ways. Radiation (photons) interacts for the last time with the surrounding matter at this epoch and carries with it the map of the structure of the universe at this epoch. This radiation cools with the expansion of the universe and has now cooled to about 2.7 degrees kelvin. This is the cosmic microwave background radiation that was discovered in 1964 (described later in this chapter). The second matter-dominated era is inaugurated after radiation and matter have decoupled. This is our era as galaxies formed and life began a few billion years afterwards. The lumpy universe observed today was formed during this second matter-dominated era by a process or processes entirely orchestrated by gravity.

Alpher and Herman first calculated the current temperature of the decoupled radiation and obtained a value of 5 degrees. The calculation of the temperature and flux of this radiation went through a number of iterations in the next 10 years. Remarkably it did not cause any stir in the community of observational astronomers. This is even more surprising as in 1940 two Australian astronomers, Andrew McKeller and Walter Adams, had measured the temperature of this radiation through the rotational transition lines of the cyanogen molecule. These astronomers determined that the excitation temperature was about 2.3 degrees. However, they did not interpret this as cosmic background radiation as the big-bang model had not yet been developed.

In 1964 Arno Penzias and Robert Wilson (researchers at the Bell Telephone Laboratories in Holmdel, New Jersey) measured noise level in their horn antenna, noise that was contaminating communication with the newly launched *Echo 1* satellite. They noticed that they were picking up sizeable amounts of signal at about 7.35 centimetres. This signal was independent of the direction, time of the day or season. The signal also did not appear to originate in the Milky Way; they ascertained this by pointing their antenna at the Andromeda galaxy, a galaxy very similar to our galaxy. It was obvious that the radiation was coming from a volume much larger than the Milky Way. Penzias and Wilson found that the equivalent temperature of the radiation they had detected was about 3.5 degrees. The meaning of this microwave noise began to be clarified as the news of these observations spread along the physics 'grape vine'. At Princeton (not far from Holmdel) James Peebles was working on a model of the early universe not unlike the model of Gamow, Alpher and Herman. He had shown that in the early universe there must have been intense high-temperature radiation. Without this radiation most of the hydrogen would have 'cooked' to heavier elements, contrary to observations. He also predicted that this radiation would now have cooled down and its current temperature would be about 10 degrees. Dicke and Peebles were planning an experiment to detect this radiation. They were too late. The observational paper of Penzias and Wilson and the theoretical paper of Dicke and Peebles and their collaborators were published jointly in the *Astrophysical Journal Letters*. Penzias and Wilson were awarded the physics Nobel Prize in 1978 for their discovery of the 'cosmic microwave background radiation'. The discovery of this radiation started a new era in cosmology. This radiation was the 'obvious relic of a superdense state of the universe' which Hoyle, the proponent of the steady-state universe, had demanded in 1955.

Physicists immediately appreciated the importance of microwave background radiation. They realised that clues to the early evolution of the universe would be buried in this radiation. There was also the intriguing possibility of testing Mach's principle. This suggestion was first made by the British astrophysicist Dennis Sciama of Cambridge University. Sciama suggested that it might be possible to measure the rotation of the local inertial frame against the cosmic background radiation. The American astrophysicist Peebles suggested that the background radiation could share some of the qualities of the ether which Michelson and Morley had failed to detect. Peebles argued that since the cosmic background radiation permeated all space it could mimic the absolute space envisioned by Newton but *without* violating Einstein's special theory of relativity. The cosmic microwave radiation is isotropic, that is the temperature is the same (within limits of measurement) in all directions. If a body moves through this radiation then the temperature in the direction of motion would appear slightly higher than the temperature in the backwash. The difference in temperature is proportional to the speed of motion relative to the speed of light. Sciama and Peebles argued that because the Milky Way is rotating (the Sun and Earth are moving at about 250 kilometres per second through the background radiation) there should be a difference of about 0.08% between the temperature of the radiation in the direction of motion and that in the backwash. In the 1970s a number of experiments were performed, mostly in the United States, to detect this anisotropy. These experiments were flown on either balloons or aircraft. The observations detected the rotation of the Milky Way and also established that the Milky Way was moving in the direction of the constellation of Leo at an extraordinarily high speed of about 600 kilometres per second. This high velocity suggested that there was a gigantic mass somewhere out there that was dragging the Milky Way under the influence of gravity. This was against the prevailing theories, according to which matter was distributed fairly homogeneously in the cosmos. Large lumps of matter were not 'expected' to exist. This was not the first time a large peculiar velocity of a galaxy had been observed. Almost 25 years previously Vera Rubin (more about her later) had reported on departures from the Hubble flow – and had been received with almost arctic coolness. In the 1980s an international collaboration discovered 'the Great Attractor' dragging the Milky Way and the

Local Group of galaxies towards the constellation of Leo. Surveys of galaxies have now revealed that the Great Attractor may not be unique and there are probably many more such large structures in the universe. Galaxies are aggregated into clusters and superclusters, like a cosmic foam; the walls of the bubbles are concentrations of galaxies while the interiors of the bubbles are vast regions of empty space.

The balloon and aircraft observations of the 1970s established that the universe was not rotating and it was expanding with remarkably uniform speed in all directions. There was no sign of asymmetry. The early events in the universe, even the big bang, appeared to be finely tuned. The picture of the universe that emerged in the early 1980s was very different from previously held beliefs. The 'new' universe was filled with radiation, which was, as far as could be determined, completely smooth. Embedded in this radiation were billions of galaxies aggregated in immense clusters and superclusters. By the late 1980s this picture was beginning to be an embarrassment. If the universe was perfectly smooth about 300 000 years after the big bang, when radiation decoupled from matter, then there was not enough time for galaxies to form and clump together about 100 million years later. The massive conglomeration of galaxies observed in the universe today must have grown from cosmic seeds present at the earliest epochs of the universe. These seeds should have left evidence of their presence in the cosmic background radiation – that is, small fluctuations in temperature, which would represent regions of slightly differing density. Under the influence of gravitation these regions of differing density would have grown into embryonic galaxies, clusters of galaxies and superclusters. Unfortunately no fluctuations in the temperature of the background radiation had been detected. Either the cosmological theories were wrong or more sensitive measurements were required.

Against this background came the announcement in April 1992 by the COBE (Cosmic Background Explorer) satellite team: they had measured fluctuations in the cosmic microwave background radiation at a level of $\pm 1.0 \times 10^{-5}$ degrees. Cosmologists and astronomers generally heaved a collective sigh of relief. The 'seeds' of galaxies and clus-

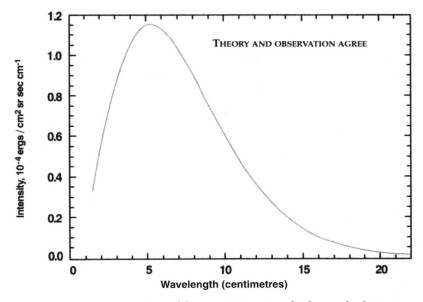

FIGURE 7.4 Spectrum of the cosmic microwave background radiation. The curve is the spectrum of a black body of 2.726 degrees. The FIRAS data were taken at 34 equally spaced points on this curve. The data match the curve with a precision of 0.03%, the error in each data point being less than the width of the curve. (Image provided by D. Leisawitz, NASA, GSFC.)

ters of galaxies might have been found. The COBE satellite had been launched in November 1989 to measure the spectrum of the microwave background radiation and the putative fluctuations in the temperature of this background. This was a collaborative programme between a number of American universities and research institutes. The first results were announced in January 1990 at the American Astronomical Society meeting. These were results from the Far Infrared Absolute Spectrophotometer (FIRAS) and these data showed that the spectrum of the background radiation was a *perfect* black-body spectrum of a *single* temperature (Figure 7.4). It is rare in science for theoretical results to agree exactly with experimental measurements and yet that is true of the spectrum of the cosmic background radiation. It is this spectrum, a spectrum of a black body of just one temperature, which lends the strongest support to the big-bang model. It tells us that the universe was once so dense and compact that it was a single body, a body that could be characterised by a single temperature.

It is now realised that the observations of anisotropy of cosmic microwave background radiation provide one of the very few means available at present of probing the universe prior to the epoch of nucleosynthesis – an epoch about which we have very little observational knowledge (gravitational wave astronomy, when possible, will provide information of much earlier epochs). For example, we do not know if the large-scale uniformity of the universe was a result of rapid expansion during inflation. Nor do we know for certain if the large structures in the universe evolved from fluctuations generated during the inflationary phase. COBE observed the fluctuations in the background radiation with a sensitivity of about 1×10^{-5} and an angular resolution of about 7 degrees. This angular resolution is very coarse; models of structure formation in the universe suggest that clusters and superclusters of galaxies observed in the present universe formed from fluctuations with angular sizes of about 1 degree. Thus higher angular resolution and higher-sensitivity maps of the cosmic microwave background radiation are essential for deeper understanding of the early universe. Various ground-based and balloon-borne experiments have been performed or are under way since the publication of the COBE results. The most ambitious programmes to map the background radiation are the PLANCK satellite selected by the European Space Agency and the MAP satellite of NASA. These satellites will map the microwave background at a number of frequencies with a sensitivity of about 2×10^{-6} and an angular resolution of 1 degree. The scientific goals of these missions are to determine fundamental cosmological parameters like the density of baryonic matter, the cosmological constant, the Hubble constant and the neutrino content of the universe. More generally these satellite missions will attempt to establish the relationship between the primordial irregularities and the large-scale structures of galaxies observed in the sky. The determination of the geometry and content of the universe by measurement of microwave background anisotropy has been given the highest priority in various national and international scientific programmes. In the first decade of the twenty-first century the PLANCK and MAP missions will attempt to achieve these goals.

The idea that the universe originated in a singular event some finite time ago, the big-bang idea, has been considerably revised and developed since its inception in the early 1930s. The incorporation of elementary particle physics in the basic big-bang idea has enhanced greatly the 'intellectual muscle' of the theory and also increased its predictive power. Only one serious competitor to the big-bang theory has emerged since the 1930s, and this was the steady-state theory developed in the late 1940s by Fred Hoyle, Thomas Gold and Herman Bondi of Cambridge University. In this theory matter in the universe is not conserved but is created continuously. Thus the matter density in the universe remains constant in spite of the observed expansion. The steady-state hypothesis evoked much interest for some time and still has some die-hard adherents attempting to produce various refinements of the original theory. The discovery of cosmic background radiation and particularly the unique temperature of this radiation are now accepted as proofs that the universe has indeed passed through a compact high-density phase. But the conditions at the singularity, at the moment of formation, have been a troubling uncertainty. Despite the considerable development of this model since the 1930s this uncertainty remains.

GALAXIES, QUASARS AND CLUSTERS OF GALAXIES

Galaxies are the basic building blocks of the universe, and they also cluster into the most massive gravitationally bound structures in the universe. In the late eighteenth century William Herschel showed that the Milky Way, which appears to the naked eye as a diffuse band of light, was a slab of stars in which our solar system is embedded. In 1784 the French astronomer Charles Messier complied a catalogue of 109 bright 'nebulous' objects. Messier was interested in comets, which also appear nebulous but which move across the sky. Unlike the comets, Messier's 109 nebulae did not move and Messier compiled his catalogue so that the comet hunters of the eighteenth century would not waste their time on these 'uninteresting objects'. At about the time Messier was compiling his catalogue, William Herschel and later his son John embarked on a far more exhaustive survey of nebulae both in the northern and southern skies. They catalogued 7840 objects, and this catalogue was later extended and published by J. L. E. Dreyer of Armagh Observatory in Northern Ireland. The nature of these nebulae varied enormously and they were given various descriptive names like diffuse nebulae, planetary nebulae and spiral nebulae. Some of these nebulae are sites of recent star formation, one of the best known examples being the Orion Nebula (M42, i.e. object 42 in Messier's catalogue). The nature of spiral nebulae was a burning issue in the first quarter of the twentieth century – were they with us (in the Milky Way) or without us? This led to the now famous debate in 1920 in the US National Academy of Sciences between two American astronomers, Harlow Shapley and Heber Curtis. Shapley was for placing the spiral nebulae within the Milky Way and Curtis for without. The majority of the assembled astronomers voted for Shapley. There was only one way to settle this issue – measure the distance to the nebulae. Hubble's observations of Cepheid variable stars in spiral nebulae established that a large number of nebulae were galaxies, far from the Milky Way in agreement with Curtis. This just goes to show that science does not progress by consensus!

The beautiful galaxies and large structures that we observe in the universe would not have formed without slight irregularities in the primordial matter at the epoch of decoupling of matter from radiation. Without these irregularities the expansion of the smooth universe would have continued in an unglamorous way. The irregularities allow gravity to 'overcome' the irresistible forces of expansion. Gravity on its own prevents the formation of very large structures and favours the formation of small-scale structures, clouds of stellar mass. However, gravity does not act alone – the hot gas exerts pressure resisting the attractive force of gravity and prevents the collapse to small structures. A happy compromise between these two forces leads to the formation of clouds of about a million solar masses. The formation of galaxies from these clouds is not yet fully understood. One possibility is that a cloud collapses into a single very massive star that quickly evolves to a black hole. These black holes act as nuclei in whose gravitational field other black holes, stars and gas are trapped to form galaxies by processes as yet unknown or unidentified. Another possibility is that a cloud collapses into globules. These globules collide and produce shock waves, which radiate away their energy of motion. This encourages the globules to merge. Also the random motion of globules and any rotational motion of the collapsing cloud or the globules force the cloud to settle into a disc (to conserve angular momentum) forming a galaxy. Stars that form before the formation of the disc will have random orbits in the halo of the galaxy and stars that form in the disc will have the disc configuration. The actual formation of galaxies may be a combination of these two processes.

In 1925, Hubble showed that galaxies could be arranged in a linear sequence. Very broadly this sequence goes thus: elliptical, spiral and irregular galaxies. The elliptical galaxies have very smooth profiles with practically no evidence of dust or gas (less than 0.01% of their mass). These galaxies have spheroidal shape with masses between 10^{13} to 10^7 solar masses. These are self-gravitating systems in which gravitational collapse is prevented by the motion of the stars. Spiral galaxies have a disc shape with a central bulge and the relative sizes of the disc and bulge vary from galaxy to galaxy. The masses of these galaxies vary from 10^{12} to 10^7 solar masses. The Milky Way is a spiral galaxy and has a mass of about 10^{11} solar masses. Very young hot stars, gas and dust, the ingredients for future star formation, define the arms of spiral galaxies. These stars, gas and dust are supported against gravitational collapse by the centripetal force due to the motion around the centre. There is ongoing star formation in the disc of these galaxies. The irregular galaxies are generally less massive than the elliptical or spiral galaxies. They have no well defined shape but have a large amount of gas and dust and there is ongoing star formation in these galaxies.

The morphological sequence of Hubble can be understood in terms of the efficiency and speed with which galaxies convert gas into stars. In evolutionary terms the elliptical galaxies are most advanced. In these galaxies star formation appears to have started about 15 billion years ago and proceeded very efficiently and speedily. In spiral galaxies star formation appears to have proceeded more slowly and less efficiently and has continued to the present day. The irregulars, like the nearby Magellanic Clouds, appear to have been most tardy at star formation. These galaxies appear to retain a large fraction (almost 30%) of their primordial gas. Although most galaxies are elliptical or spiral, some have a highly disturbed appearance. These galaxies have interacted gravitationally with other galaxies. The shapes and structures of these interacting galaxies are determined by the tides raised by the gravitational interaction. This disturbance in gas can create strong shock waves, which can trigger star formation.

In 1942 the Pacific War was in full swing and Los Angeles had been blacked out (for the expected air attack by the Japanese). This was an astronomer's dream. Karl Seyfert, an experienced astronomer at the Mount Wilson Observatory, decided to take advantage of this extraordinary situation to obtain long-exposure spectra. Seyfert had noticed that a small fraction of the spiral galaxies had intense blue point-like nuclei and he decided to obtain spectra of these nuclei. These spectra had strong emission lines not unlike those produced by ionised gas. But there was a difference: these lines were not narrow and sharp like those observed from laboratory sources, but very broad. This width is now believed to be due to Doppler broadening produced by the high-speed turbulent motion of gas clouds emitting the lines. The width suggested random velocities a small fraction of the speed of light – 10 to 100 times higher than typical gas motion observed in our galaxy. The Armenian astronomer B. E. Markarian of the Byurakan Observatory in the former Soviet Armenia has discovered that some elliptical galaxies also have intense blue nuclei not unlike the spiral Seyfert galaxies. More surprises were to follow.

By the late 1940s radio astronomers were discovering very 'loud' radio sources in the sky. In 1954 Walter Baade and Rudolf Minkowski of Mount Wilson and Palomar Observatories showed that the loudest radio source, called Cygnus A, was associated with a faint galaxy with a redshift of 0.05, implying that this galaxy is about 300 times further away than the Andromeda galaxy. The image of the galaxy had a double structure which suggested, wrongly as it turned out, that these were two colliding galaxies. Further radio observations showed that the radio emission from Cygnus A is not from the galaxy itself but from two lobes symmetrically placed about the galaxy. Clearly this galaxy is not just a collection of stars like the general run of the mill galaxies. There is something very odd going on in this and similar radio-loud galaxies. In the late 1950s the 3rd Cambridge Radio Survey was completed. Two sources in this survey, 3C48 and 3C273, were bright enough to be observed optically but their positions were not known with sufficient accuracy to attempt spectroscopic studies. In 1962 Cyril Hazard and his colleagues, taking advantage of the occultation by the Moon, used the Parks Radio Telescope in Australia to measure the position and the apparent size of 3C273. They showed that the source size was less than 0.5 arcseconds – that is, it was 'point-like' and not extended like a galaxy. In February 1963 Maarten Schmidt, at the Palomar Observatory, used the very accurate position determined by the Parks observations to obtain a spectrum of 3C273 using the 5-metre Hale telescope. The spectrum completely baffled Schmidt; he had not seen anything like this before. He eventually noticed that the relative wavelengths of the strong lines fitted the spectrum of hydrogen but with a redshift of 0.158 – this source was nearly 2 billion light-years away, if it was receding because of cosmic expansion. But another surprise was in store. The 'point-like' size (now confirmed by observations of the very rapid fluctuations in brightness of this source, particularly at X-ray energies) suggested that the physical size of 3C273 is just a few light-minutes or smaller than the size of the solar system. Because these objects appeared stellar they were dubbed quasars or QSO (for quasi-stellar objects). This was a cultural shock for the astronomy establishment. These sources had been discovered by a technique that was still in its infancy and not by the tried and tested optical observations. And, if the cosmic distances of these objects were to be believed, then they were emitting a thousand billion times the energy emitted by the Sun and from a region no bigger than the solar system. These objects are a million times more luminous than the Milky Way. Even a faint quasar is almost 100 times more luminous than the Milky Way. How was this vast luminosity generated? In 1963 Fred Hoyle and William A. Fowler suggested that the enormous energy released by quasars was a result of gravitational collapse. Discovery of quasars was one of the factors that triggered the renewed interest in gravitational physics in the early 1960s. The radio lobes of radio galaxies posed a similar energy problem. It was realised that the radio emission from the lobes was synchrotron radiation emitted by electrons at near-light-speed spiralling around magnetic field lines. Geoffrey Burbidge at the University of California in San Diego (USA) showed that the energy in the magnetic field and the spiralling electrons was comparable to the energy released by total annihilation of about ten million suns.

A zoo of active galactic nuclei (AGNs) has now been discovered and about 1% of observed galaxies have an active nucleus. Quasars of redshift as high as 5 have now been discovered – these objects were formed when the universe was less than 20% of its age, but the peak of quasar formation appears to have been at a redshift of about 2.5. AGNs emit radiation over the entire electromagnetic spectrum, from gamma rays to radio. These galaxies confront us with two problems: what is the source of the enormous luminosity and what powers the radio jets (and gamma-ray emission)? By early 1960 astrophysicists had worked out in some detail the nuclear fusion reactions powering the stars (described later in this chapter). The efficiency of nuclear power reactions is about 1%; that is, 1% of available mass is converted to energy. (The efficiency of chemical reactions, that is, the usable energy relative to the total chemical energy available in the fuel, is only about 1 part in 10^8. It is this small fraction of energy that is the power generated in the internal combustion engine of cars and aeroplanes). Thus about a billion solar masses of nuclear fuel would be required to power a low-luminosity AGN, and all this fuel would have to be consumed, no loss or wastage is allowed. The amount of fuel required for higher-luminosity AGNs could be truly staggering. The energy requirement is only a problem if AGNs are at their cosmic distances as implied by their observed redshifts. In the late 1960s and early 1970s there was a lively debate on this issue. It was argued that the Hubble Law, which works so well for normal galaxies, was not applicable to such unusual objects as AGNs. However, the 'parent' galaxy of a number of AGNs has now been observed and the Hubble Law is certainly valid for these galaxies. The consensus of opinion now is that AGNs are indeed at the cosmic distances indicated by their redshifts. Implosion of a normal star leading to a supernova explosion has a mass to power efficiency of about 10%. Thus about 100 million normal stars would provide enough energy to power a low-luminosity AGN. Also, the emitted radiation would be variable because of explosions. It is possible that the low-luminosity AGNs are powered in this way. However, for moderate and high-luminosity AGNs the number of exploding stars required and the rate at which they would have to go supernova become unrealistically large. By the late 1970s it was accepted that the source of power for AGNs was gravity, the primary process being the accretion of mass onto a giant black hole. The black hole in an AGN forms (by processes as yet unidentified) by accretion and in turn accretes more mass from the surrounding space. This matter comes from the interstellar gas from the surrounding galaxy, gas captured in a gravitational encounter with another galaxy or gas released when a star passing close to the black hole is disrupted by gravitational tidal forces. The incoming gas swirls in towards the black hole to conserve angular momentum and spirals into regions of even stronger gravity. This compresses the gas, which heats and ionises it and the gas begins to radiate energy. The gas eventually settles into a giant accretion disc. Bits of matter eventually fall into the black hole releasing the gravitational potential energy. The energy released is several per cent of the rest mass of the in-falling mass. This process may be similar to that occurring in X-ray binaries, discussed later.

The gravitational energy released by the matter accreting (falling) on the black hole cannot produce the relativistic jets observed in radio galaxies. To produce these jets it is necessary to tap the energy of the rotating black hole directly. Roger Penrose of Oxford University first postulated this possibility and Roger Blandford and Roman Znajek of Cambridge University worked out the details for AGNs. The spinning black hole drags with it the space around it (gravitomagnetic effect) which spins the magnetic field lines threaded through the hole. Hot ionised gas (plasma) is propelled along these spinning field lines and the plasma shoots out along the axis of the spinning hole creating two jets. Electrons in the plasma gyrate along the field lines emitting the observed radio waves. Low-energy photons from the hot accretion disc are boosted to gamma-ray energies as they bounce off these high-energy electrons. Thus the power in the jets (and the radio lobes) comes directly from the black hole's enormous rotational energy.

Although giant black holes appear to be the most likely powerhouse for AGNs, the formation of these powerhouses is still shrouded in mystery. The nuclei of galaxies, including that of our Milky Way, are obscured by thick clouds of dust and gas and are invisible at optical wavelengths. But observations at infrared (and radio and gamma-ray) wavelengths have revealed that this region is very tightly packed with stars. In the vicinity of the Sun there are about 0.006 stars per cubic light-year, but in the nucleus of our galaxy there are almost 2 stars per cubic light-year. This is also true of the nuclei of other galaxies. These stars are moving at speeds of several hundred kilometres per second under gravitational attraction. The fate of this concentration of fast-moving stars is not well understood partly because the gravitational encounter of two stars is not well understood. When stars are far away (a few hundred stellar radii say) they can be treated as point masses, as Newton had done to investigate the motion of planets around the Sun. But this approximation is not valid when stars approach each other to within a few stellar radii. At these distances the gravitational tidal force deforms their structure and also slows down the stars by drag forces. If the approach speeds and distances are just right, the stars can become locked as a binary system, called a tidal capture binary. If the velocities of the stars drop below the escape speed of each star, the stars spiral together and eventually coalesce into a single massive star. These massive stars evolve rapidly and explode as supernovae, scattering gas in the nucleus and creating neutron stars or black holes. At higher relative speeds the stars smack into each other, shatter and scatter the gas in the nucleus. Gravity causes this gas to accumulate in the centre. The fate of these massive stars, gas, neutron stars and black holes, packed in the nucleus, has not been worked out in detail but the gut feeling is that the end product is a black hole of millions, if not billions, of solar masses.

Apart from the circumstantial evidence of the spectacular activity in the nuclei of active galaxies there is very little observational evidence for the presence of massive black holes there. Black holes can only reveal their presence through their gravitational interaction with the surrounding material. The one galactic nucleus that we can observe in detail is the nucleus of the Milky Way, which is about 32 616 light years away. The intervening shroud of gas and dust prevents observations at optical and ultraviolet wavelengths but the nuclear region has been mapped at radio, infrared and gamma-ray wavelengths. This region is very complex; a dense cluster of stars is embedded in filaments and blobs of gas. The centre of this mess is a point-like source of radio emission called Sagittarius A* (Sgr A*). Could this be a million-solar-mass black hole? The compact size and the lack of transverse motion support arguments for a massive object. Recently Andreas Eckart and Reinhard Genzel of the Max Planck Institute for Extraterrestrial Physics in Garching, Germany, have determined the full three-dimensional velocity structure of about 200 stars in the vicinity of Sgr A*. For these observations they used a specially built infrared camera and the 3.5-metre New Technology Telescope of the European Southern Observatory in Chile. These data enabled the two astronomers to determine the orbits of these stars in the gravitational potential of Sgr A*. These studies conclusively show that stars in the central cluster follow a Keplerian velocity distribution around a black hole of 2.6 million solar masses.

Direct evidence of a massive black hole at the centre of other galaxies is not easy to obtain. However, other pieces of observational evidence indicate that there is a concentration of mass in the nuclei of a number of galaxies. The extra gravity of a massive black hole alters the motion of stars in the nucleus of a galaxy and this distortion of orbits causes the stars to move closer to the black hole. This concentration of stars in the sphere of influence of the hole causes an enhancement of starlight at the nucleus of the galaxy. Such enhancements of light have been observed at the centre of a number of galaxies. However, this is not an unambiguous way of establishing the presence of a black hole in a galaxy as the blip of light could be due to accretion of matter on the central massive object, which may or may not be a black hole. The stars in the central concentration move in the gravitational potential of the central massive object and the stars closest to the centre move anomalously fast. Unambiguous observations of the motion of stars in this central concentration of stars have only been possible with the Hubble Space Telescope. These observations have revealed a disc of glowing gas orbiting the centre of M87, a weakly active giant elliptical galaxy with weak extended radio lobes and an optical jet. This disc is about 60 light-years from the centre. Doppler shift of spectral lines emitted by gas in this disc combined with Kepler's law suggest a massive central object of 2 to 3 billion solar mass. Similar massive black holes have now been discovered in a number of nearby galaxies. Our nearest large neighbour, the Andromeda galaxy or M31, appears to have a black hole of 30 to 70 million solar masses. Much more precise and compelling evidence of a black hole has been obtained from radio interferometric observations of water vapour maser lines in the galaxy NGC4258. The spatial and spectral resolution of a radio interferometer is almost a factor of a hundred better than that of HST. Whereas HST can map structures over a scale of tens of light-years, NGC4258 has been mapped down to a scale of one light-year. The evidence for concentration of mass in this galaxy is correspondingly stronger and it unambiguously points to a single massive black hole in the galaxy. Mass estimates of black holes in the nuclei of about 15 nearby (normal) galaxies have now been made and it is likely that the majority of luminous galaxies contain a dormant black hole in their nucleus.

The angular resolution of HST and radio observations limits observations of gas to about 100 000 gravitational radii (a gravitational radius is defined as GM/c^2 where M is the mass of the central object; the gravitational radius of the Earth is about 4.3 millimetres and that of the Sun is about 1.5 kilometres). The inner region of an accretion disc can be probed at X-ray energies. The gas in the inner region of an accretion disc is several million degrees hot and it is highly ionised and emits radiation at X-ray energies. Most of the X-ray radiation emerges from within about 20 gravitational radii from a black hole. Spectral lines emitted by this hot gas provide a powerful tool to study this inner region. Matter close to the black hole moves very fast – typical speeds are of the order of 100 000 kilometres per second (the speed of the gas detected by the HST and radio interferometers is about 1000 kilometres per second). There is also a velocity gradient in this matter; gas close to the black hole moves faster than the gas further out. If the gravitational field is assumed to be purely Newtonian then the lines emitted by the hot gas will be symmetrically broadened by the Doppler effect. However, the velocity of gas in the inner regions is so high that Newtonian gravity is no longer applicable and relativity has to be considered. Special relativistic effects cause the light from the approaching side of the disc to be beamed and this radiation appears brighter than the radiation from the receding side. Also because of special relativity the radiation from the inner region is redshifted relative to radiation from the outer region of the disc. General relativistic effects introduce further (gravitational) redshift in the radiation emitted from the inner regions of the disc. Thus an emission line from this inner region is very broad and skewed with two peaks: the blue peak is almost at the rest wavelength of the line and is brighter than the red peak. The red peak is spread out in a wide tail. The receding and the innermost parts of the disc emit this red tail. The strongest line emitted by the hot gas from the accretion disc is that of ionised iron. Observations of this line have been possible with the launch, in February 1993, of the Japanese–US X-ray satellite ASCA. A number of AGNs have now been observed with ASCA and the highly asymmetric line has been detected (an example is shown Figure 7.5) in all of these spectra. These

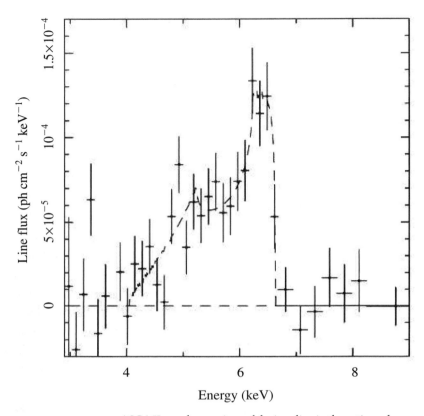

FIGURE 7.5 ASCA X-ray observations of the iron line in the active galaxy MCG-6-30-15. The continuous curve is a theoretical profile of a line emitted from gas rotating in a high gravitational field. The low energy is the red wing of the line and the blue wing is the high-energy side. (Figure provided by A. Fabian and reproduced with permission of Y. Tanaka.)

observations are consistent with, if not confirmation of, the theories of accretion around a massive black hole in AGNs. The highly distorted shape of the line supports the belief that the emitting gas is in a strong gravitational field and this region is best described by general relativity.

DARK MATTER

A lot is now known about the morphology, star formation, chemical evolution and stellar dynamics in galaxies, but the problem of the formation of galaxies is still unresolved. Galaxies formed at an early and remote epoch. Deep infrared observations seem to suggest that the peak of galaxy formation was when the universe was 3 to 5 billion years old but the galaxies may have started to form when the universe was only about a billion years old. The exotic conditions in the universe at these early epochs are not known at present, but these conditions must have allowed, by processes as yet unidentified, gravitationally bound gas clouds to condense out of the primordial gas. These seeds and the conditions then prevailing must have determined the rich variety of galaxies (certainly the non-interacting galaxies) now observed. An additional problem in understanding the formation of galaxies is the lack of knowledge of the actual mass of a galaxy. Only about 10% of the measured mass of a galaxy is observable matter, the remaining 90% is currently invisible. The nature of this dark matter is not known but it is this matter which determines the gravitational field of a galaxy and which, therefore, must have played a crucial part in the formation of the galaxy. A considerable body of observational evidence for dark matter has been accumulated over the past few decades.

In 1927 the Dutch astronomer Jan Hendrik Oort (1900–1992) measured the velocity of stars perpendicular to the plane of the Milky Way. Oort was one of the most important figures in the twentieth-century effort to understand the nature of the Milky Way galaxy. He was educated at the University of Groningen, joined the Leiden Observatory in 1924 and became its director in 1945, a position he held until 1970. From radio observations Oort determined that the Sun was 32616 light-years from the centre of the galaxy and took 225 million years to complete an orbit around it. From his 1927 measurements of the velocity of stars perpendicular to the plane of the Milky Way Oort estimated the total amount of gravitating matter in the galactic disc. He deduced that the mass density in the vicinity of the Sun was about 0.15 solar mass per square parsec or about 3.0×10^{-4} kilograms per square metre. This is the amount of material in the plane of the galaxy in a column that has a cross-section of one square metre and height equal to the thickness of the galaxy. To Oort's surprise this was about double the amount of material locked up in the stars, gas and dust in the plane of the galaxy. Various refinements in the measurement of the perpendicular component of stellar velocity and estimates of the mass in stars, gas and dust, made since Oort, suggest that the conclusion Oort reached was correct: there is gravitating material in the plane of the Milky Way which is not visible. This is the origin of the local hidden mass problem. One possibility is that this mass is locked in very low-mass stars, stars of such low mass that they cannot become hydrogen-burning main sequence stars. These are called *brown dwarfs* to reflect their uncertain status. These stars are very cool; their only source of energy may be the gravitational binding energy, which they radiate over a very long period. Heroic attempts have been made to detect these low-mass stars and determine their contribution to the local mass density.

But this is not the end of the story of dark matter in the Milky Way or in other galaxies. In the early 1970s Jeremiah Ostriker and James Peebles at Princeton University analysed in detail the stability of galaxies. Ostriker and Peebles showed that the gravitational pull of the material in the stars, dust and gas in the galaxies was not sufficient to hold a galaxy together. The stars should fly apart! They concluded that the galaxies could only be 'held together' if the stars, dust and gas were surrounded by a massive but invisible halo. This halo had to be massive enough to contain almost 90% of the actual mass of a galaxy. This suggestion was far too 'adventurous' for the 1970s and the work of Ostriker and Peebles was ignored.

Another person who suspected that 'what we see is not what there is' was Fritz Zwicky, back in the 1930s. He was studying the Coma cluster of galaxies, which is about 300 million light-years away. The velocities of the galaxies in the cluster puzzled Zwicky. They were too high. At the measured velocities the galaxies in the cluster should fly apart and the cluster should evaporate. Clearly the cluster was held together by the gravitating mass in the cluster and Zwicky decided to measure this mass. He first estimated the amount of mass in the stars, gas and dust in the galaxies; this is the visible mass. He then estimated the dynamic mass of the cluster from the rotational speed of the galaxies in the cluster and an application of Newton's laws of motion. Zwicky found that the dynamical mass was almost a factor of 20 higher than the visible mass. He reported that there must be 'dark matter' (he was the first astronomer to use this term) holding the cluster together. But this was the 1930s, it was only a few years since Hubble had demonstrated that the galaxies were well beyond the Milky Way. Extragalactic astronomy was still a very young subject. Moreover, Zwicky was a difficult person and the consensus among astronomers was that Zwicky's results were 'premature'. Today the mass of the X-ray emitting hot gas in the cluster, which was discovered after the launch of X-ray satellites, should be added to the visible matter. There is, nevertheless, a factor of 10 discrepancy between the amount of visible matter and the total gravitational matter. Almost 90% of matter in a cluster announces its presence only through its gravitational influence.

The credit for putting dark matter on the astronomical agenda goes to Vera Rubin and her colleagues at the Carnegie Institution in Washington DC. This is perhaps one of the most shameful tales, in modern times, of discrimination against women scientists. Vera Rubin became interested in astronomy when she was just 10 years old. Her father, who built her a telescope and took her to meetings of amateur astronomers, encouraged her interest in sciences. Her teachers advised her to 'stay away from science' and tried to steer her towards more 'lady-like' subjects. But Rubin was not to be discouraged so easily and in 1948 she graduated from Vassar College, New York, in science. After graduation Rubin applied to the prestigious Princeton University for post-graduate studies in astronomy, but was not accepted. Princeton did not accept women graduate students in astronomy until 1971! Disappointed, Rubin decided to go to Cornell University for her post-graduate studies, where her husband, a physical chemist, had moved. At Cornell, Rubin studied physics under Hans Bethe and Richard Feynman, both Nobel laureates. Her master's thesis demonstrated that the motion of some galaxies deviates from the uniform expansion of the simple big-bang theory. Her conclusion was met by the full wrath of male hostility and her paper was rejected for publication. In 1954 Rubin received her doctorate from Georgetown University in Washington (her supervisor was George Gamow) where her family had moved earlier.

It was not realised at the time, but Rubin's work was a landmark. Disappointed at the controversy her work was causing, Rubin turned to what appeared to her (and others) at the time a mundane area of astronomy, the measurement of rotation of galaxies. Unknowingly, she had opened a Pandora's Box. She began measurements of the rotation of the nearest (and visible to the naked eye) giant galaxy, the Andromeda galaxy. She expected the outer regions of the galaxy to rotate more slowly than the inner regions. This is what is expected from Newtonian mechanics and what is seen in the solar system. In the solar system the planets orbiting the large central mass (the Sun) have decreasing orbital speeds as a function of orbital radius; the planet Mercury travels at about 172 163 kilometres per hour while the outer planet Pluto moves at about 16 895 kilometres per hour. To her utter amazement, Rubin discovered that the velocity of the gas in the galaxy was constant, it did not matter whether the gas was close to the centre or at the outer rim, it moved at same speed. At first she thought that there was something peculiar about the Andromeda galaxy. But Rubin and her colleagues at the Carnegie Institution have now measured the velocity of over 200 galaxies, and in every case the velocity from the centre to the edge of the galaxy is nearly constant. The speed of very tenuous gas well beyond the visible extent of a galaxy has now been measured with the aid of the narrow line of atomic hydrogen. The velocity stays constant. These observations demonstrate that the gas in a galaxy is experiencing the gravitational attraction of a mass that is considerably higher than the visible mass in the galaxy. It suggests that the galaxy is rotating like a solid body and the stars (and gas) are not rotating around a central mass. The rotation curves of galaxies is conclusive demonstration of invisible matter in galaxies and this evidence is now overwhelming. For this pioneering work Vera Rubin was elected to the National Academy of Science in 1981. Rubin's courage and determination are an inspiration to all established and aspiring women scientists. Unfortunately the science establishment, still-male-dominated, has not learned much. Women are still discriminated against in sciences generally and in access to senior academic and industrial positions particularly.

The American astronomer Dennis Zaritsky took the method of measuring stellar velocities to infer the mass of a galaxy enclosed by the orbit of the star, pioneered by Rubin, to its logical conclusion. He measured the orbital velocity of the satellite galaxies (the Magellanic Clouds being the best known) in orbit round the Milky Way, to infer the mass enclosed by the orbits of these galaxies. The most distant satellite galaxy, Leo I, is 700 000 light years from the galactic centre, 20 times as far out as the Sun. The orbit of Leo I encloses a very large volume. In 1998 Zaritsky reported that the halo of the Milky Way extends out to Leo I, and has a mass of a thousand billion suns, 10 times greater than all the visible mass in the Milky Way. Zaritsky has since shown that other galaxies also have similar huge halos.

The dark matter in galaxies, which Rubin had discovered, manifests itself only through its gravitational effect. There is no other way to deduce the presence of this material. In 1986 Bohdan Paczynski, of Princeton University, realised that the gravity of the dark matter can be used to detect lumps of this matter. These lumps of dark matter could be in the form of very dark stars known as MAssive Compact Halo Objects, or MACHOs. A MACHO would gravitationally lens the light of a background star and magnify the image of the star as it passed in front of the star. This magnification of starlight would be short-lived and it would also be symmetric, that is, the rate of brightening and fading would be equal. Also the brightening and fading would be similar in different colours because gravitational lenses are achromatic. The time scale of a lensing event is proportional to the square root of the mass of the lensing object. Thus the sudden brightening of a star, with these characteristic signatures, would reveal the presence of dark matter in the foreground. A number of research groups have undertaken the search for MACHOs in the halo of our Galaxy, using this technique. In 1994 two groups independently reported the predicted stellar brightening and a number of similar events have now been detected. The unambiguous interpretation of these data has proved difficult and at present it is not established that these lensing events are due to clumps of dark matter. The lensing is more likely to be due to 'normal' stars. At present there is little evidence for a large population of MACHOs in the halo of our Galaxy.

The existence of dark matter is no longer in doubt, but the nature of the dark matter is still unknown. An obvious hypothesis is that this matter is 'baryonic' – that is, the stuff the everyday world is made of. This baryonic dark matter is not observed because it may be hidden in an undiscovered population of Jupiter-size planets, neutron stars, black holes or remnants of a hypothetical population of very massive stars, which may have formed in the early history of galaxies. However, the big-bang theory of the universe sets an upper limit on the total amount of baryonic matter in the universe. This limit is set by the amount of helium and deuterium (and lithium) observed today. In the context of the big-bang theory, to create the observed amount of helium and deuterium the total baryonic matter has to be less than the total amount of dark matter observed. Thus it seems unlikely that the dark matter in the huge galactic halos and in large astronomical structures like clusters and superclusters is a form of baryonic matter.

These arguments have encouraged elementary particle physicists to unleash their imagination. It has been suggested that most of the dark matter in the universe may be 'nonbaryonic' – fundamental particles not found in familiar matter. Unfortunately there is a stable of (mostly hypothetical) particles to choose from. The most likely candidate is the neutrino, which has the advantage over other nonbaryonic candidates in that it is known to exist. Wolfgang Pauli predicted this particle in 1930, to preserve the principle of conservation of energy in the radioactive decay of nuclei. In 1933 the great Italian-American physicist Enrico Fermi published the first comprehensive theory of this particle, which he christened the neutrino ('little neutral one' in Italian). This work was considered so speculative that his paper was rejected by the British journal *Nature*. Neutrinos are produced in copious numbers in fission reactions in nuclear reactors and this is where they were first detected in 1953. The neutrino has been shown to have no charge and is thoroughly antisocial, refusing to interact with anything. According to the big-bang theory the universe should be flooded with neutrinos, outnumbering baryons by a factor of about 10^8. The neutrino as a likely candidate for dark matter was first suggested in the 1970s. Unfortunately the mass of the neutrino was not known then. Developments in particle physics in the 1980s indicated that there may be three types of neutrinos and they might not be massless after all. If the mass of a neutrino is about 10^{-4} times the mass of the electron then there would be enough neutrino mass in the universe to account for the observed dark matter. Analysis of the burst of neutrinos from the Supernova 1987A, in the Large Magellanic Cloud, suggests that the mass of the electron neutrino may be less than 4×10^{-4} times the mass of the electron. Firmer measurements of the neutrino mass may be forthcoming in the near future from the reactor neutrino beam experiments. Neutrinos however have one disadvantage as a dark matter candidate: they move too fast to be captured in the gravitational potential well of galaxies. Another candidate for dark matter is the family of heavy neutral particles known as Weakly Interacting Massive Particles (WIMPs). In the past decade the search for dark matter has shifted from the sky to the Earth. There are now over a dozen experiments, located in deep mines, searching for WIMPs and other exotic particles conjectured by particle physicists. The incentive for these extremely complex experiments comes not just from a desire to detect the dark matter but (more importantly) from the enormous contribution they would make to our understanding of the nature of matter and nuclear forces. Almost 90% of gravitating material in galaxies and clusters is this putative dark matter and the formation of these astronomical structures must depend on the nature of this matter. It is a sobering thought that at present we only know that this matter exists.

Can the grip of gravity halt the expansion of the universe? To answer this question it is necessary to determine the gravitational pull the universe has on itself and this depends on the amount of matter in the universe and the putative cosmological constant. The average density of matter just enough to resist the expansion of the universe is known as the critical density. The numerical value of this density is about 2×10^{-26} kg m^{-3} (kilograms per cubic metre) – about five atoms per cubic metre. Note that this is the average matter density – that is, the density when all matter in all galaxies, clusters of galaxies, stars and planets is smeared over the entire universe. If the density of matter in the universe is greater than this critical density, then gravity will pull the universe upon itself, slowing down the current expansion and eventually reversing it. This is known as the closed universe. If the total density is less than the critical density then the universe will continue to expand forever; this is known as an open universe. The amount of all matter in the universe – normal matter in stars, galaxies and clusters of galaxies plus the invisible dark matter in the galaxies and clusters – appears to be about 30% of the critical density and this is unlikely to create enough gravity to stop the expansion of the universe. It is possible that a large quantity of dark matter in the universe has not been detected yet. But it should be emphasised that the detection of dark matter in the galaxies and clusters of galaxies does not imply that there is additional dark matter in the universe to achieve the closure density.

The fate of the universe may not be determined only by the amount of matter in the universe. If the cosmological constant has a finite value, then this will also resist the pull of gravity and will have a role to play in the evolution of the universe. Observations of distant supernovae, published in 1998, indicate that these supernovae are dimmer then those expected in a universe with no cosmological constant. This suggests that the universe was expanding, at the time the light was emitted, at a rate faster than that expected from the mean mass density of matter in the universe. This acceleration in cosmic expansion seems to be due to the repulsive effect of the cosmological constant overcoming the attractive effect of gravitating matter. These data, combined with the curvature of space determined by comparing the angular scale of structures seen in the cosmic microwave background, suggest that the universe is spatially 'flat', i.e. space is Euclidean with zero curvature. About 70% of the energy density necessary to make the universe flat appears to arise from the cosmological constant, the matter density providing the other 30%. These observations pose an intriguing problem: what is the origin of the repulsive energy of the cosmological constant and why is the repulsive energy density so similar to that of the matter density? These questions cannot be answered with the available supernovae observations. More data on supernovae at higher redshifts and more rigorous analysis of these data are required, as are more detailed observations of the cosmic background radiation, before unambiguous conclusions can be reached about the total matter density in the universe and the nature of the cosmological constant.

THE BIRTH, LIFE AND DEATH OF STARS

The life history of a star is determined by the balance between gravity and heat (generated by nuclear reactions). Gravity has a natural tendency to contract a star by gravitational attraction, and heat (and resulting pressure) tends to expand the star. The balance between these two forces accounts for astronomy's most exotic objects – white dwarfs, neutron stars, black holes and supernovae.

In the formation of stars the spiral structure of a galaxy plays a critical role. The gas in the arms is highly ionised, tenuous and hot. The high pressure in this gas prevents gravity from compressing the gas. But the density of the gas is also high and there are frequent collisions of atoms and ions with dust grains, which cool the gas. In some locations the gas is cooled sufficiently for hydrogen atoms to combine to hydrogen molecules. The dynamical processes in the arms cause the molecular gas to concentrate into giant molecular clouds. These clouds can be about a hundred light-years across and can have a mass of a few hundred thousand solar masses[4]. The mass of these clouds increases (by accretion of more gas and other clouds) slowly with time and when the mass and dimensions reach a certain critical limit (called the Jeans mass and Jeans length respectively, after the British physicist James Jeans) the cloud becomes gravitationally unstable and begins to contract. The contracting cloud fragments into smaller and denser clouds of between a thousand and ten thousand solar masses. These smaller clouds continue to collapse, but the central region of the cloud collapses faster than the outer regions. The clouds also rotate and, as they contract, the spin of the cloud increases to conserve angular momentum. This rotation leads to the formation of an accretion disc around the compressed central region. As the gravitational compression increases the temperature of the central protostar rises and at some stage the protostar becomes hot enough for nuclear reactions to start burning hydrogen into helium. The gas and dust in the accretion disc eventually form into planets and moons around the new star (described later in this chapter). The nuclear burning produces a powerful wind of hot gas, which drives away the gas in the cocoon around the young star, and the star becomes visible at optical wavelengths.

About 25% of the gas and dust in the original giant molecular cloud is converted into stars. The new stars cover a large range in mass, from a fraction of a solar mass to about 100 solar masses. This large cluster of stars is held together by the gravitational pull of the residual gas. But the radiation and wind from the massive protostars heat the remaining gas and drive it out of the cluster. Consequently, the mass of the cluster decreases to a level where the grip of gravity on the stars is reduced and the stars escape the cluster. The few open clusters seen in the sky are in the process of evaporating as they have lost their placental gas and the gravitational attraction on the stars has slackened. It should be emphasised that this is only a outline of star formation; the details of star formation are not well understood. For example, it is not known why rotation and magnetic fields in a molecular cloud do not resist the gravitational collapse of the cloud. Detailed observations made with the Infrared Space Observatory (which was launched by ESA in 1995 and obtained data for about 24 months before the liquid helium used to cool the on-board telescope and detectors was exhausted) may shed light on some of the highly complex processes. These studies have implications beyond the immediate question of star formation. For example, the evolution of galaxies can only be understood if the rate of star formation under different astrophysical conditions is known.

Towards the end of the nineteenth century and the first quarter of the twentieth century two important questions were challenging scientists. What is the source of energy emitted by stars? How are chemical elements of our world created? Behind these two questions can be seen the footprints of the two major developments of this period, namely the theories of gravitation and quantum mechanics. The energy emitted by the Sun was particularly intriguing. According to the theory of William Thomson (later Lord Kelvin) and Hermann von Helmholtz, the source of solar energy was gravitation. These scientists reasoned that an imperceptible contraction of the Sun would account for the emitted energy. However, this theory predicted that the age of Sun should be no more than about one hundred million years. This was in sharp disagreement with the much larger age of the Earth postulated by evolutionary biologists and geologists. This led to the famous dispute over the age of the Earth. Lord Kelvin attacked what appeared to him the virtually indefinite time scale of geologists like Charles Lyell and biologists like Charles Darwin. The consensus of opinion favoured the physically based time scale, largely owing to the authority of Lord Kelvin and physics generally. By 1890 the evidence accumulated by geologists made it increasingly difficult to accept the short time scale favoured by the physicists.

The problem of the source of energy emitted by the stars was (one of the many problems) occupying the fertile mind of the British physicist Sir Arthur Stanley Eddington (1882–1944). Eddington did pioneering work on stellar structure, the relation between stellar mass and luminosity, relativity and many other areas of physics and astrophysics. He was born in a Quaker family in Kendal near Lake Windermere in the north-west of England. In 1898 he went to Owens College in Manchester and in 1902 to Trinity College, Cambridge. At Trinity he won every mathematical honour including Senior Wrangler and was appointed Plumian Professor of astronomy and director of Cambridge Observatory in 1913–1914. In a seminal book (*Stellar Movement and the Structure of the Universe*) published in 1914 he summarised his mathematically elegant investigation of dynamics of globular stellar systems. He also proposed the then unusual thesis that the spiral nebulae were galaxies like the Milky Way. It was another 15 years before Hubble was able to prove this by measuring distances to galaxies. Eddington was the first 'interpreter' of the theory of relativity in the English language. His great thesis *The Mathematical Theory of Relativity* published in 1923 was described by Einstein as the finest presentation of the subject in any language. Eddington also wrote extensively to popularise the subjects in which he was a recognised expert. A deeply religious man, Eddington had strongly held Quaker beliefs and maintained that the nature of the universe could not be discovered through science alone and must be sought through spiritual belief. Towards the end of his life he attempted to unify quantum mechanics and general relativity. He believed that through this unification it would be possible to calculate values of fundamental universal constants such as the number of atoms in the universe and the ratio of mass of the proton to that of the electron. Eddington did not complete this vast synthesis but what was achieved was published in 1946 in the *Fundamental Theory*.

In 1917 Eddington postulated energy generation in stars, to avoid the short time scale of the stellar contraction theory. In the following 10 years he suggested a number of processes to account for this internal energy, all based on the then little understood nuclear physics. In the early 1920s even the composition of stars was not known. It was generally assumed that stars had approximately the same composition as the Earth, with iron being the most abundant element. In 1925, Cecilia Payne (later Payne-Gaposchkin), one of the very few women to make a significant impact on science in the early part of twentieth century, determined the composition of stars from spectroscopic data. She concluded that the principal constituents of the stellar atmosphere were hydrogen and helium. This idea was against the prevailing view and the celebrated American astronomer Henry Norris Russell persuaded Payne to discount her observations and conclude instead that her results were 'almost certainly not real'. However, some years later Payne's original result was vindicated and it was established that hydrogen was indeed the most abundant element in stars. Remarkably, Eddington had conjectured this.

In his book *The Internal Constitution of Stars*, published in 1929, Eddington concluded that the 'manifestations of the greatest bodies in the universe are linked to those of the smallest'. In the early 1930s nuclear physicists had started to turn their attention to problems in astronomy. Inspired by experiments in laboratories, they believed that what was possible in a laboratory might also be possible in the interior of stars. In 1938 Hans Albrecht Bethe and Charles Critchfield proposed the first quantitative model of solar (and stellar) energy production. Bethe (1906–) was born in Strasbourg (then in the German Empire) and studied in Munich, Germany, under the pioneer quantum mechanist Arnold Sommerfeld. He obtained his doctorate in 1928 and by the early 1930s was recognised for his expertise in quantum physics. Being half-Jewish he fled Germany in 1933, first to England and from there to the United States. He settled at Cornell University in Ithaca, New York, and soon established himself as the foremost authority in nuclear theory. Bethe and Critchfield identified several nuclear reactions that keep the stars shining. The main chain reaction (known as the proton–proton cycle) is the conversion of hydrogen to helium through three nuclear reactions:

$$^1H + {}^1H \rightarrow {}^2D + e^+ + \nu, {}^2D + {}^1H \rightarrow {}^3He + \gamma,$$
$$^3He + {}^3He \rightarrow {}^4He + 2\,{}^1H$$

where ^1H is a hydrogen nucleus (or a proton), ^2D is deuterium, e^+ is positron, γ is a gamma-ray photon, ν is a neutrino, ^3He is an isotope of helium and ^4He is helium. The net result of this sequence of reactions is:

$$4\,{}^1H \rightarrow {}^4He + 2e^+ + 2\nu + 2\gamma$$

that is, conversion of four hydrogen atoms (or protons) to a helium atom. (The neutrino had not been established in 1938 but is now accepted as an essential diagnostic tool for this reaction.) The total mass of four protons is 4.03252 atomic units and that of the helium nucleus is 4.00386. There is thus a difference of 0.02866 units of mass or 5×10^{-29} kilograms in the above reaction. In the stellar interior this excess mass is converted to energy (through Einstein's $E = mc^2$) and is equal to 4×10^{-12} joules. This reaction accounts for about 90% of the energy emitted by stars (it makes the stars 'shine') on the main sequence – that is, the large majority of stars. The luminosity (the total power output) of the Sun is 3.8×10^{26} watts, equivalent to 9.5×10^{37} proton–proton reactions per second and a mass loss of 1.5×10^{17} kilograms per year. This is only about 10^{-14} of the total mass of the Sun, so there is plenty of 'fuel' to keep the Sun shining for few billion years. In 1967 Bethe was awarded the Nobel Prize in physics for his work on nuclear reactions and energy production in stars.

To start the p–p reaction the temperature in the star has to be high; this is achieved by gravitational self-contraction of the star. The internal temperature a star can attain is determined by the mass of the star. The theory of stellar structure suggests that the lower mass limit for hydrogen burning to start is about 0.08 solar mass and the minimum mass for a cloud to undergo gravitational collapse is 0.01 solar mass. For comparison the mass of Jupiter is 0.001 solar mass. The search for very faint low-mass stars is thus a crucial part of the study of formation and evolution of stars. It is important to identify the amount of material in a molecular cloud that is converted to low-mass stars and to see if the theoretical criterion for the onset of fusion is correct.

It is impossible to observe the evolution of individual stars, since a star's life is considerably longer than that of an astronomer. The reason stellar evolution can be studied in great detail is because the luminosity and the surface temperature of stars occupy a very well defined region of the temperature–luminosity diagram. This is known as the *Hertzsprung–Russell* (or H–R) diagram after the Danish astronomer Ejnar Hertzsprung and the American Henry Norris Russell; they discovered the H–R diagram in 1914. A schematic representation of this diagram is shown in Figure 7.6. Stars that begin to burn hydrogen to helium lie along a sequence that runs from bottom right to top left. This is known as the *main sequence* and most of the known stars lie along this sequence. The high-mass stars at the top left have a surface temperature of about 30 000 degrees and are almost 10^5 times

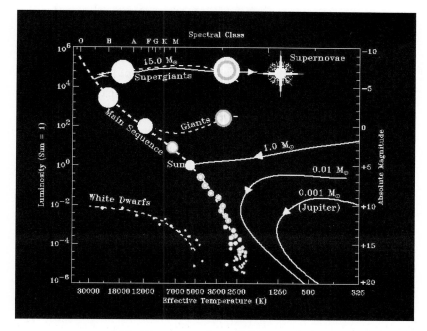

FIGURE 7.6 The Hertzsprung–Russell diagram. The luminosity (which is proportional to the mass) of a star is plotted as a function of the surface temperature. As a star contracts out of a molecular gas cloud it moves in from the right and arrives on the main sequence by the time of ignition of nuclear reactions in the core. The star stays on the main sequence for most of the rest of its life. The actual length of time on the main sequence is determined by the mass of the star; a high-mass star can burn its supply of fuel in a few hundred million years and evolve away (to the right) from the main sequence; a low-mass star, like the Sun, can stay on the main sequence for billions of years. As stars evolve from the main sequence they can either explode as supernovae or condense to white dwarfs. (Figure provided by B. Kellett.)

as luminous as the Sun. The surface temperature of the lower-mass stars, at the bottom right of the main sequence, is about 3000 degrees. The luminosity of these stars is only about one-hundredth of the Sun. The Sun lies at about the middle of the main sequence and its surface temperature is about 6000 degrees; this is a very ordinary star.

The life and evolution of a star are determined by its mass. Ironically, lower-mass stars live longer than their higher-mass siblings. This is because the lower-mass stars have cooler cores and so the nuclear reactions in their core, which are very sensitive to temperature, are slower. On the other hand, in the high-temperature core of high-mass stars the reactions are fast and the available fuel (hydrogen) is consumed at a high rate. The average life of high-mass stars (those with masses greater than about 10 solar masses) is about 10 million years, but that of low- and intermediate-mass (i.e. lower than 10 solar masses) stars is a few billion years. Thus many generations of high-mass stars have come and gone since the formation of the universe, but all low-mass stars ever formed are still around.

A star of low or intermediate mass spends between 80 and 90% of its life on the main sequence. When the star has converted about 12% of its hydrogen into helium, a point known as the *Schonberg–Chandrasekhar* limit, the star becomes unstable. The core of such a star cools because most of the hydrogen has been converted to helium and the next set of nuclear reactions has not started. The cold core contracts because it cannot resist the attractive force of gravity. The envelope, on the other hand, continues to burn hydrogen and it expands. It can expand to between 100 and 1000 times its diameter on the main sequence. As the expanding outer layer cools the emitted spectrum of the star shifts towards the red and the star 'becomes' a *red giant*. A red giant has a very dense core and a very large tenuous outer region. The density of gas in this outer region is lower than that of the Earth's atmosphere. Compared with the main sequence stars, the red giants are rather rare but they are very bright and stand out quite conspicuously in the sky. The two well known red giants in our galaxy are Betelgeuse in the constellation of Orion and Antares in the constellation Scorpio. The Greek name Antares means 'the rival of Mars' because of the brilliant red colour of this star.

The gravitational contraction of the core of a red giant eventually raises the temperature of the core high enough to fuse helium nuclei into those of carbon and oxygen. These nuclear reactions produce about half of the carbon in the universe and a large fraction of elements like tin, cadmium and lead. Dust grains also form in the relatively cool outer envelope and are expelled in the galaxy. Dust grains are very efficient at cooling molecular gas clouds to trigger the formation of a new generation of stars and planets. The length of the helium-burning phase is about one-quarter of the main sequence lifetime. The star continues to shine through hydrogen burning in the envelope and this hydrogen burning sustains the envelope against gravitational collapse. The star eventually expels the outer hydrogen-rich atmosphere. The remnant star collapses rapidly under gravity and its surface temperature rises to about 100 000 degrees. At this high temperature the star radiates copiously at ultraviolet and soft X-ray wavelengths. Ultimately the star ceases to burn hydrogen and evolves into a cooling *white dwarf*. In a white dwarf the quantum mechanical pressure of the electrons in the star balances the gravitational force compressing the star. The gravitational pressure determines the final size of a white dwarf and so a white dwarf gets smaller as its mass increases. Typically the size of a white dwarf is comparable to that of the Earth. It is estimated that there are about 10 billion white dwarfs in our galaxy but very few of these are visible from Earth. Companions of both Sirius and Procyon are white dwarfs. A white dwarf does not produce any energy as the nuclear reactions in its interior have stopped and so has the gravitational contraction. It continues to shine because of the very high temperature of the surface gas. As time passes, an isolated white dwarf gradually cools and fades.

High-mass stars (stars between 11 and 50 solar masses) follow a very different life cycle. After a relatively brief time on the main sequence, a single star also forms a carbon core. But, unlike the core of the low- and intermediate-mass stars, the mass of the core of the high-mass star is large enough for it to continue to contract under gravitational attraction. A series of nuclear reactions follows, ending with a core of ironlike elements. Further reactions are not possible, as these would require input of energy. Lighter elements continue to burn in the outer shell of the star, adding mass to the core. If the core mass exceeds 1.4 solar masses (the *Chandrasekhar* limit) the core begins to collapse under gravitational contraction. In the contracting core the iron-like nuclei decompose into helium nuclei, which in turn fragment into neutrons under the relentless gravitational contraction. Eventually the collapsing core forms a *neutron star*, which resists the gravitational contraction by quantum mechanical repulsion of neutrons – similar to quantum mechanical pressure of electrons in white dwarfs. A neutron star of 1.5 to 2 solar masses would be about 25 kilometres across. The catastrophic core collapse that forms the neutron star lasts for about 0.1 seconds and almost all of the gravitational potential energy of the collapsing star is converted into neutrinos. Most of the neutrinos fly unimpeded through the remnant star at almost the speed of light. A very small fraction collide with the atomic nuclei in the dying star and impart enough energy to the nuclei to blow off the outer layers of the star – the star becomes a *Type II supernova*. In the exploding supernova some of the gravitational potential energy of the star is converted to heat, light and motion of gas. The supernova shines for several weeks with a luminosity of 10 to 100 billion suns and can outshine the parent galaxy. If the collapse of the star is asymmetric then in the last phase of the collapse the supernova also loses energy by emitting gravitational radiation. In future it will be possible to observe this radiation and investigate the terminal moments of a star's death.

If a star on the main sequence has a very high mass, say between 50 and 100 solar masses, the star loses mass via stellar wind at a very high rate, and when the star leaves the main sequence it is left with only a helium core. The evolutionary path of these stars is qualitatively similar to that of the lower-mass stars but the mass of the iron core of such a star is so high that the quantum mechanical pressure of neutrons cannot halt the gravitational collapse. Instead, the collapse continues until the star's gravity prevents the escape of light – and the star becomes a *black hole* (described in detail later in this chapter). The explosion which creates a black hole is called a *Type Ib supernova*. Part of the gravitational energy released during the collapse and formation of a black hole is radiated as gravitational waves. The isolated black hole created in this supernova is visible only for a limited time as the leftover stellar debris falls into the gravitational potential well of the hole and becomes hot. Thereafter, the black hole can only become visible if it gravitationally captures an interstellar cloud or if it lenses the light of a background star.

The supernova explosions, Type II and Type Ib, blow the heavy elements formed in the precursor star into the interstellar medium. These heavy elements are important for our existence and for the formation of the next generation of stars. The blast from the explosion sweeps up the gas in the interstellar space and compresses it into dense clouds. When the mass of a cloud is large enough gravity takes over, the cloud is compressed further, and it condenses into a new star with planets around it, which may support life.

Supernovae are rare events. Only three supernovae have been recorded in the Milky Way in the past 1000 years. In 1054 Chinese astronomers recorded a 'guest star' that was brighter than Venus and was visible in daytime. The remnant of this star is the Crab nebula, which has been intensively studied in the past few decades. The next two supernovae occurred in quick succession in 1572 and 1604 and, as has been narrated earlier, were observed as bright new stars by Tycho Brahe and Johannes Kepler respectively. A galactic supernova has not been observed since 1604 but a bright supernova was observed in our nearest neighbour galaxy, the Large Magellanic Cloud (LMC), on 23 February 1987. The full might of late twentieth-century astronomy was deployed to study this supernova. The position of the supernova coincided with the position of a bright blue 15 solar-mass star designated Sanduleak −69 202 (this designation means that the star was catalogued by an astronomer called Sanduleak and this was star number 202 in a band around declination of −69 degrees). The explosion was observed with a host of ground-based and space-based instruments and over the entire electromagnetic spectrum, from gamma rays to radio. But perhaps the most important and certainly the most exciting observation was that of neutrinos released by the collapsing core. These neutrinos were detected by instruments located in the Kemioka zinc mine (north of Tokyo) in Japan and in the Morton salt mine under Lake Erie near Cleveland, Ohio in the USA. At 7:35 a.m. (Greenwich Mean Time) both experiments detected simultaneously a pulse of neutrinos, which lasted for about 10 seconds. The visible light emitted by the exploding star arrived at Earth between 9 a.m. and 11 a.m. (GMT). The delay between the neutrino pulse and the optical pulse is what is expected from the models of supernova explosions. In addition the inferred neutrino luminosity and neutrino energy spectrum also match those expected when the binding energy of a neutron star is released. The neutrino results confirm the general picture of stellar evolution outlined above, and the central role of gravitation in the life of a star. These observations herald the birth of a new branch of observational astronomy – neutrino astronomy.

The above picture of stellar evolution refers only to single stars or widely separated stars. Over 50% of stars in our galaxy are members of binary systems whose individual members are gravitationally bound to each other. If a star in a binary system is close enough to exchange material with its companion then the evolution of the star will be dramatically different from that of an isolated star. In a binary system the centripetal force associated with the motion of the binaries is added to the gravitational potential of individual stars. This results in equipotential surfaces around the pair and a critical surface, which encompasses both stars. This is known as the *Roche lobe*, after Edouard Roche (1820–1883), who analysed binary systems in the middle of the nineteenth century. The point of contact of the two lobes is known as the inner *Lagrangian point* L_1. When the stars are smaller than their Roche lobes their evolution is more or less similar to that of an isolated star. But when the more massive companion of the pair (called the primary) moves off the main sequence into a red giant phase it expands and fills its Roche lobe. Matter always seeks the lowest gravitational potential and flows through the Lagrangian point onto the secondary (Figure 7.7). Thus initially the mass of the primary decreases and that of the secondary increases. This can result in the secondary becoming massive. The secondary can now evolve into a red giant and the process of mass transfer is reversed. The end point can be a pair consisting of a high-mass star and a white dwarf, or a white dwarf binary system. In a high-

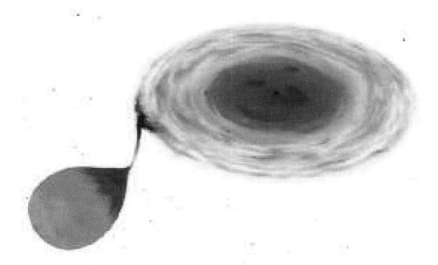

FIGURE 7.7 An artist's representation of a binary stellar system. As the massive star in the binary evolves to a red giant it can fill its Roche lobe and the gas from its outer atmosphere can escape through the inner Lagrangian point onto the surface of the companion, accelerating its evolution. Material created with support to AURA/STScI from NASA contract NAS5-26555.

mass/white-dwarf pair the mass accreted by the white dwarf can raise it above the critical mass for stability of a white dwarf. The dwarf then explodes as a *Type I supernova* liberating about 10% of the rest mass energy and forms a neutron star. Type I supernovae have remarkably uniform properties as the progenitor star in all cases is similar. If in a binary pair the less massive star becomes a supernova then the system can stay bound creating a binary system with a neutron star or a black hole. If the more massive star in the binary explodes (as a Type II supernova) then the system can disrupt, propelling the companion at high speed. This may be a plausible explanation for the high-velocity neutron stars observed in the galaxy.

Binary stars can be detected by the periodic eclipse the stars produce or the slight periodic shift in the spectral lines of one or both stars. The periods of these systems can range from a few minutes to many years. These eclipses can be modelled very precisely to estimate the stellar masses. These mass estimates have confirmed that the compact companion in a number of binary systems is either a neutron star or a black hole.

NEUTRON STARS

One autumn day in 1967, postgraduate student Jocelyn Bell, at the Mullard Radio Astronomy Observatory of Cambridge University, was startled to pick up regular radio signals from the sky. The conclusion seemed obvious – extraterrestrials were attempting to contact Earth. However, Jocelyn Bell and her supervisor Anthony Hewish dismissed this as fanciful and settled for a mundane but astronomically plausible explanation. They surmised, correctly, that the radio signals were emitted by a star, but a star with rather unusual characteristics. The duration of the pulses suggested a very rapidly rotating small star. But a star rotating as fast as the pulses suggested must have an immense gravitational field, to prevent the star from flying apart. The Cambridge astronomers concluded that the pulses were emitted by a rotating neutron star. The first pulsar had been discovered. In their initial survey Hewish and Bell discovered three more pulsars. In 1974 Hewish was awarded the Nobel Prize in physics for this discovery, sharing this prize with Sir Martin Ryle, one of the pioneers of radio astronomy. The decision of the Swedish Academy not to include Jocelyn Bell in the award is incomprehensible.

The Swiss physicist Fritz Zwicky (1898–1974) had predicted neutron stars long before they were discovered. In 1925 Zwicky had been lured to the California Institute of Technology, California, USA to do theoretical research in atomic physics but he was increasingly drawn to astrophysics. Today he is regarded as a genius but during his lifetime he was often dismissed as a maverick. He was a difficult person often in conflict with his colleagues and his publications often included violent attacks on them. In 1930s he proposed neutron stars to explain supernovae and cosmic rays – the most energetic phenomenon then known to astronomers and physicists. Zwicky claimed that the neutron stars were the end product of evolution of all massive stars. That is, light stars end up as white dwarfs and massive stars as neutron stars. This was very comforting for astronomers and physicists of the 1930s as it avoided resorting to the incomprehensible and unacceptable black holes. Einstein and Eddington heaved a sigh of relief (see next section – *Black holes*). Disappointingly, the models of neutron stars suggested that the only detectable emission from these stars was the surface thermal radiation. In a prescient paper, published in 1967 but before the discovery of pulsars, the Italian astronomer Franco Pacini predicted that neutron stars might be observable at radio wavelengths if they were magnetised and were oblique rotators. A similar prediction had also been made, at about this time, by the American astrophysicist Thomas Gold. In the autumn of that year Zwicky's neutron stars were broadcasting their own story.

Pulsars have now been conclusively identified as rotating magnetised neutron stars. The radio radiation is produced by electrons accelerated in the magnetic field at the magnetic poles of the star; radio pulses are detected when the beam of radiation swipes past an observer, not unlike a beam from a lighthouse. The period of these pulses is remarkably stable. In a recently discovered pulsar with a period of about 1 millisecond the period is stable to better than one part in 10^{13} over one year. This makes pulsars very accurate clocks. Pulsars provided the first evidence of highly compact objects, objects with central density as high as 10^{18} kilograms per cubic metre and radius as small as 10 kilometres. Orbital periods of binary systems with neutron stars suggest that most neutron stars have a mass between 1.4 and 1.5 solar masses, in very good agreement with theory. In a neutron star the gravitational field is so high that general relativity is not a small correction, as in the solar system, but plays a crucial role. This makes the neutron stars ideal laboratories to study matter in a strong gravitational field. Because their behaviour can vary over observable time scales, they can be rich sources of information not only on general relativity but also about nuclear physics and astrophysics. As described earlier, the binary pulsars provide, at present, the only evidence of gravitational radiation.

It is estimated that about 1% of the stars in our galaxy are neutron stars and over a thousand of these have now been discovered. As a radio pulsar radiates away its rotational energy, its spin frequency decreases. The rate of change of a star's rotation is a fundamental observable parameter. From these measurements the enormous magnetic field of neutron stars has been measured. Small variations in the precise timing of radio pulsars have also led to the astonishing discovery of multiple Earth-mass planets orbiting some neutron stars. Although most neutron stars have been discovered as radio pulsars, only a small fraction of the radiated energy actually goes into radio emission. The bulk of the energy is radiated as high-energy (X-ray and gamma-ray) photons. The depth of a neutron star's gravitational potential well makes accretion of material another prime energy source for the star. The brightest accreting neutron stars reside in binary systems and accrete matter from their companion, either by tidally stripping material from their surface or gathering up some of the wind they expel. These accreting neutron stars have luminosities more than a thousand times that of the Sun. Most of this energy is emitted at X-ray wavelengths, and observations with the X-ray and gamma-ray observatories launched in the last 10 years have done much to unravel the secrets of these highly compact objects and the gravitational and nuclear processes taking place on and around them.

Initially a neutron star cools mainly by emitting neutrinos. These neutrinos are emitted when the protons and electrons in the star are squeezed together to neutrons by gravity. As the supply of free protons and electrons runs out the neutrino emission gradually diminishes and the star cools mainly by radiating X-rays from its surface. The inside of a neutron star is an uncharted territory. In this region gravity distorts matter to a degree where the conventional nuclear physics is no longer applicable. Some theorists have speculated that inside some neutron stars gravity may be strong enough to set free the quarks which make up the neutrons. It will be possible to study the conditions inside a neutron star as the technology of neutrino astronomy develops. Changes in the highly stable spin rate of a neutron star may also carry information about the interior of the neutron stars. An entirely new window on neutron stars will be opened when the ground- and space-based gravitational wave observatories come on-line (see Chapter 6, Dicke). With these observatories it will be possible to detect gravitational waves from orbiting neutron stars and also detect the coalescence of neutron star binary systems. Thus in the not too distant future X-rays, gamma rays, neutrinos and gravity waves will provide new details of neutron stars.

BLACK HOLES

A neutron star is the last known stable form of a star. In more compact and massive objects the gravitational attraction can overcome the quantum mechanical pressure holding up a neutron star and the object collapses to a black hole. The American physicist John Archibald Wheeler coined the name 'black hole' to indicate that even radiation cannot escape from this object. But the escape of light from the surface of a star was first considered by the Reverend John Michell, rector of Thornhill in Yorkshire, England. In a paper presented to the Royal Society (by Henry Cavendish) on 27 November 1783 Michell states[5]:

> if the semi-diameter of sphære of the same density with the sun were to exceed that of the sun in the proportion of 500 to 1, a body falling from an infinite height towards it, would have acquired at its surface a greater velocity than that of light, and consequently, supposing light to be attracted by the same force in proportion to its vis inertiæ, with other bodies, all light emitted from such a body would be made to return towards it, by its own proper gravity

Michell used Newtonian theory of gravity and Newton's concept of the corpuscle nature of light to compute a critical size of a star for which the escape velocity would equal the speed of light. A light corpuscle emitted from such a star would rise off the surface of the star but its speed would gradually decrease and the corpuscle would ultimately fall back to the star's surface (Figure 7.8). Michell speculated that the universe was full of such invisible massive stars. Thirteen years later the French physicist and astronomer Pierre Simon Laplace repeated this speculation in the first edition of his *Le Systeme du Monde* (but he did not refer to Michell's work). The idea was abandoned soon afterwards because Newton's corpuscular theory gave way to the wave theory of light formulated by Christiaan Huygens.

Over 200 years had to pass before scientists returned to the problem of the effect of gravity on light. Karl Schwarzschild took the first step. His solution of Einstein's equations of general relativity indicated that the curvature of space-time increased around massive compact objects. The solutions also indicated that there was a critical circumference and if the star's surface coincided with this circumference then the curvature of space-time around the star would be so high that light emitted from the surface would not be able to escape the stellar surface. In modern terms the star's strong gravity creates a black hole horizon around the star. The result of the 'Schwarzschild' description of a compact star is essentially similar to that of Michell and Laplace – a star as small as the critical circumference will be invisible to a distant observer, that is, it will be a black hole. But that is where the similarity ends. In the treatment of Michell and Laplace the velocity of light decreases as it attempts to overcome the gravity of the compact star. In this treatment space and time are absolute but the speed of light is variable. In the Schwarzschild description the speed of light is absolute and space and time are relative. Around a compact object space-time is curved to prevent light escaping the star but the speed of light does not change.

In the 1920s and 1930s both Einstein and Eddington had profound misgivings about black holes. They both felt that 'they knew' how nature should behave and black holes did not fit into their intuition, particularly Eddington's. Eddington firmly believed that 'there should be a law of Nature to prevent a star from behaving in this absurd way!' This was to prove traumatic for the young Indian astrophysicist Subrahmanyan Chandrasekhar (1910–1995). Chandrasekhar was born in Lahore (in modern-day Pakistan) in a gifted family (in 1912 his uncle C. V. Raman was awarded the Nobel Prize in physics). His early university education was at the Presidency College in Madras, India, and he published his first scientific paper at the age of 18. He was introduced to the new statistics of Fermi and Dirac and the application of the statistics to the electron gas in metals by the great German physicist Arnold Sommerfeld, during the latter's visit to the Presidency College in 1928. In 1930 Chandrasekhar went to the University of Cambridge in England to pursue his postgraduate studies. During the 18-day voyage from Madras to Southampton, he decided to 'look into' the terminal stages of stellar evolution as described in Eddington's book *The Internal Constitution of Stars* and the quantum mechanical pressure which prevented the gravitational collapse of white dwarfs. The British physicist R. H. Fowler (who was to be Chandrasekhar's supervisor at Cambridge) had demonstrated this. Chandrasekhar was investigating the stability of white dwarf stars and he discovered (as Fowler had done before him) that the speed of electrons in the high-density gas of the stellar atmosphere was comparable to the speed of light. At these speeds the use of general relativity was mandatory, a fact which he, and Fowler before him, had ignored. So Chandrasekhar went 'back to the blackboard' and recomputed the properties of high-density, high-speed gas under relativistic conditions. The result was astonishing: if the mass of the high-density star exceeded 1.4 solar masses then it would have difficulty supporting itself against the squeeze of gravity. This meant that no white dwarf should have mass greater than about 1.4 solar masses. Chandrasekhar attempted to get his result published in Britain but Fowler declined to forward his paper for publication in a British scientific journal as he (and the famous British astronomer E. A. Milne) failed to understand Chandrasekhar's reasoning. Chandrasekhar eventually managed to get this paper published in an American journal where it languished, completely ignored by the astronomical community.

Chandrasekhar returned to white dwarfs three years later, after completing his doctorate and this time investigated the stability of stars for a whole range of parameters defining the internal structure of stars. The modelled results were in good agreement with the few measured values of mass and radius of white dwarfs and the maximum mass of 1.4 solar masses appeared to be confirmed. He presented his results to the Royal Astronomical Society in London on 11 January 1935. Eddington gave a talk immediately after that of Chandrasekhar. Eddington was perhaps one of the very few people at that meeting who understood the implications of Chandrasekhar's work. He must have realised that a star more massive than 1.4 solar masses must implode under gravitational pressure – that is, it would became a black hole. And this Eddington found unacceptable. He dismissed Chandrasekhar's work and concluded, wrongly as it turns out, that the absurd result was a consequence of inadequate meshing of special relativity and quantum mechanics. Eddington was an 'establishment figure', his achievements were legendary, and the assembled fellows of the Royal Astronomical Society sycophantically fell in behind him and dismissed Chandrasekhar's work. There was an edge of cruelty in Eddington's behaviour; both Eddington and Chandrasekhar were in Trinity College at Cambridge and Eddington had had a number of opportunities to discuss with Chandrasekhar the problems of synthesising special relativity and quantum mechanics. He had known that Chandrasekhar was going to present the work on white dwarf stability at the meeting on 11 January but failed to discuss any perceived problems, choosing instead to humiliate the young Indian scientist publicly. Chandrasekhar was shocked and disappointed. Leading scientists privately supported him but publicly stayed behind Eddington. The celebrated Danish atomic physicist Niels Bohr agreed with Chandrasekhar but did not publicly oppose Eddington as he 'wanted to avoid a controversy'. Disheartened, Chandrasekhar abandoned his investigations of the terminal stages of stellar evolution. In 1936 he accepted a position at the University of Chicago. However, this was not the end of his travail; the dean of physical sciences at Chicago was against a 'coloured man' teaching in his department. But the Russian-born director of the Yerkes Observatory, Otto Struve, was having none of this: he wanted the best available theoretical astrophysicist and Chandrasekhar was the best. Chandrasekhar stayed at the University of Chicago for the rest of his life. In his almost 60 years at Chicago he made fundamental contributions to a number of fields in physics and astronomy and became a legend in his own time. In the early 1950s he returned to the question of white dwarfs and the terminal stages of stellar evolution. By then nobody had any doubts about the veracity of his work. For his pioneering work in relativistic astrophysics Chandrasekhar was awarded the Nobel Prize in physics in 1983. The limit of 1.4 solar masses for white dwarfs is today known as the *Chandrasekhar limit*.

Chandrasekhar's calculations had shown that stars under 1.4 solar masses would become white dwarfs. But would all stars more massive than this end up as neutron stars or was there a maximum mass limit to a neutron star above which the formation of a black hole was inevitable? Robert Oppenheimer (later to become the Director of the Manhattan atomic bomb project) and his student George Volkoff tackled this problem in 1938. They were able to show that there was a maximum allowed mass for a neutron star and this was between about half a solar mass and several solar masses. This large uncertainty was a result of the inadequate knowledge, in 1938, of nuclear forces. Today we know that the maximum allowed mass is between 1.5 and 3.0 solar masses. Oppenheimer and Volkoff published their paper on neutron stars in 1939; they pointedly failed to mention Zwicky or his publications on neutron stars. This was because of the very low opinion Oppenheimer had of Zwicky. Zwicky seethed with rage and sent off his own paper on neutron stars. This paper does not contain a single reference to the Oppenheimer–Volkoff paper that had been published just two months earlier! Contrary to popular misconception, scientists are not high-minded, always in pursuit of scientific truth. Personal likes, dislikes and jealousies play almost as big a part in scientific research as in soap operas.

The work of Chandrasekhar, Oppenheimer and Volkoff had unequivocally shown that a massive star that avoided becoming a white dwarf or a neutron star would end up as a black hole – unless, of course, there was a way of preventing this, as Einstein and Eddington expected there should be. Before this could be investigated further the Second World War followed by the Cold War intervened. These drained most of the talented minds in the United States and the former Soviet Union into weapons research. By the mid-1950s some of these scientists had started to return to universities and colleges, and to full-time teaching and research. They were also looking for challenging research topics to which they could apply the expertise and tools they had developed to fashion more and more destructive weapons. And black hole research is very challenging. The scientists who took up the challenge of black hole research were John Archibald Wheeler in the United States, Yakov Borisovich Zel'dovich and Igor Novikov in the former Soviet Union and Dennis Sciama in Britain. These four scientists and their students (and their students) along with a few 'oldies' like Chandrasekhar (who had returned to the study of black holes in 1974) have developed our current understanding of black holes.

The escape of a light photon from a contracting star can best be described with the help of the space-time diagram shown in Figure 7.8(*a*). In this diagram are plotted the trajectories of particles on the surface of a star as the star implodes. Photon A, emitted before the star implodes, will escape to infinity, moving equal distance in equal time. Photon B, emitted when the star has collapsed to about half its original radius, will be delayed in the curved space-time of the star; that is, it will be redshifted (photon A will also be redshifted by the gravity of a 'normal' star, but this will be small). Photon C, emitted when the star has contracted to the Schwarzschild radius, will not be able to escape and will stay on this critical surface. The trajectories of particles after the star reaches the Schwarzschild radius are not shown because information from radii smaller than this radius cannot be transmitted to a distant observer. This description should be compared with the 'Newtonian trajectory' of a photon emitted from a star of Schwarzschild radius and as conjectured by Michell and Laplace: this is shown in Figure 7.8(*b*). Looked at in this way the critical surface at Schwarzschild radius is an *event horizon*. An event horizon is a boundary that separates points in space-time from which photons can escape to infinity from points from which they cannot escape. An event taking place at or inside this horizon will not be observable to an outside observer. Analogously, a body falling into the black hole will appear to slow down as it approaches the black hole

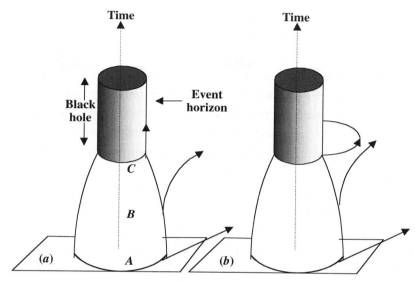

FIGURE 7.8 Space-time diagrams depicting the implosion of a star to form a black hole. Photons emitted at time A, before gravitational contraction begins, escape to infinity. Photons emitted at time B, when the star has contracted to half the original radius, also escape to infinity but are redshifted, and photons emitted at time C, when the star has contracted to the event horizon, cannot escape the event horizon in the relativistic black hole (a) but *fall* back on the star in the Newtonian description (b).

because the signal emitted by the body will take longer to reach an observer. It is impossible to determine the exact time when the body will cross the event horizon because the signals emitted at this surface will take infinitely long to reach an observer. The Schwarzschild singularity so much disliked by Einstein (and Eddington) was this event horizon. The real nature of gravitational collapse within the context of the general theory of relativity became clearer in the 1960s when the British astrophysicist Stephen Hawking and mathematician Roger Penrose proved that the fact that space-time was curved in on itself implied that in a gravitational collapse there would always be a singularity, that is, a region where the density was infinite. Einstein and Eddington would have liked this singularity even less. This is a much more sinister region, a region where matter, space and even time lose their identity. At a singularity time can have a beginning or an end: a beginning when the universe 'came into existence' in the big bang and an end when a star collapses. This singularity signals the inability of general relativity to describe situations where a gravitational field changes rapidly over small distances. It is hoped that the problems of gravitational singularity will be solved when the theory of gravity is combined with quantum theory, a theory that successfully describes phenomena occurring at small distances.

It is now accepted that black holes are not just theoretical curiosities. They can account for some of the most spectacular astronomical discoveries of recent times. Black holes represent the ultimate triumph of gravity over all other forces. Early physicists and astronomers did not regard black holes as physical realities because the mathematical treatment of Schwarzschild and also that of Oppenheimer and Volkoff was highly idealised. The imploding star which Oppenheimer and Volkoff investigated was assumed to be a nonrotating perfect sphere. Sceptics argued that nature is never so generous (real stars are oblate spheroids) and any imperfection in the stellar surface would be magnified during the collapse. Thus collapsing masses would miss each other and the formation of a singularity would be avoided. The problem of asymmetries in a collapse was not tackled till the late 1960s when Stephen Hawking and Roger Penrose showed that in a collapse Einstein's theory inevitably leads to a singularity surrounded by an event horizon. Any surface irregularity is radiated away as gravitational radiation and the black hole settles into a stable state characterised only by its mass, angular momentum and electric charge. The black hole can then reveal its presence only by its gravitational interaction with the surrounding matter. The radiation of asymmetries and the reduction of the description of a black hole to just a few parameters is colloquially known as the 'black holes have no hair' theorem. What this means is that all black holes look alike but for a few distinguishing characteristics. The collapse considered by Hawking and Penrose is also rather idealised: they did not consider the collapse of a rotating body or star and there are no nonrotating stars in nature. The implosion of a rotating star is a nontrivial problem and the details of collapse of such star are still not known. But the New Zealand physicist Roy Kerr has investigated in detail the space-time around a rotating black hole. Initially Kerr's solutions were considered to be mathematical curiosities but are now accepted as the true description of space-time around any realistic black hole.

Black holes, however, are not completely black. In 1974 Stephen Hawking investigated the behaviour of matter in the vicinity of a black hole, using quantum mechanics. He showed that a black hole can emit particles at a steady rate and the spectrum of these particles is precisely thermal. This means that a black hole creates and emits particles just as a hot body emits heat or photons (of long wavelength). Hawking showed that it was possible to parametrise this process by a temperature of the black hole and that this temperature was proportional to the surface gravity and inversely proportional to the mass of the black hole. For a black hole, with the mass of the Sun the temperature of the emitted particle would be only about one ten-millionth of degree. But this temperature would increase very rapidly as the mass of the black hole decreases.

Energy is the key to understanding black holes. The energetic processes taking place in binary systems, particularly in systems with compact companions such as a neutron star or a black hole, were discovered when these binaries were detected at X-ray energies. The first X-ray observatory, UHURU, was launched in December 1970. A number of X-ray observatories have been launched since and the X-ray observations have confirmed the general picture of the evolution of stars in binary systems. The X-ray luminosity of these binaries is produced by gas from the primary accreting on the compact secondary. The accretion can occur either because the secondary is embedded in the stellar wind of the primary or by mass transfer through the first Lagrangian point. The kinetic energy of the in-falling material is converted into thermal energy when the material hits the companion. The thermal energy released is the gravitational binding energy of the material on the surface of the compact secondary. The energy release mechanism is extremely efficient. About 10% of the rest-mass energy of the in-falling material is radiated when it impacts on a nonrotating black hole, and this increases to 43% in the case of a rotating black hole. The archetypal stellar black hole is the Cygnus X-1 binary system. The prodigiously powerful X-ray emission from this system varies so rapidly it has to be a compact object. Cygnus X-1 is a binary stellar system and the analysis of the binary period and the application of Kepler's laws suggests that the mass of the compact object in this system is 8 solar masses. This conclusively establishes the identity of the compact object as a black hole (the mass of the other two compact objects known to astrophysicists, the white dwarf and the neutron star, would be less than 3 solar masses). Such dynamical arguments have turned out to be the only possible way to decisively identify black holes, and nine black hole binary systems have now been identified. Like neutron star astrophysics, black hole astrophysics is set to make rapid progress as new X-ray and gamma-ray observatories come on-line and as data become available from the gravitational wave observatories in the future.

PLANETS AND PLANETARY SYSTEMS

The formation of planets and planetary systems is intimately linked with the formation of stars. As described above, a star is formed from the detritus of dying stars and remnants of the early universe. These collect into clouds and when the cloud accumulates a large enough mass it collapses under its own gravity. The collapsing material accumulates quickly into a central proto-star. The collapsing cloud also has sufficient angular momentum to prevent all material from spiralling into the proto-star. The exact fraction of the parent cloud that is prevented from spiralling into the proto-star is not known but it is believed to be large. The conservation of the angular momentum of the spiralling gas forces it to settle into the 'invariable plane' of the rotating system creating a large but very thin disc. The formation of this stable thin disc takes about 10^5 years after the onset of the free-fall collapse of the parent cloud – almost instantaneously in cosmic time. Infrared surveys of star-forming regions

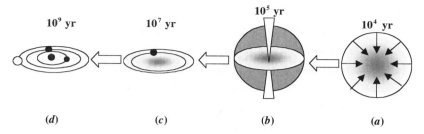

FIGURE 7.9 The main stages of formation of a planetary system: (a) The collapse of an interstellar cloud. (b) The active disc and outflow phase. (c) The passive disc and planet formation phase. (d) The formation of a stable planetary system as the star reaches the main sequence.

conducted in the 1980s have shown that at least half of all newly formed single stars possess a proto-planetary disc. These observations provide direct support for the eighteenth-century ideas of Kant and Laplace that our own planetary system might have formed from such a disc. A proto-planetary disc is composed mostly of gas but also of small particles or dust grains. As the proto-star evolves the dust grains settle into a dense layer in the mid-plane of the disc and begin to stick together as they collide. During the next 10^4-10^5 years large rocks and small asteroids grow from these dust particles. As the gravitational attraction of the larger asteroids grows they attract neighbouring pebbles and rocks and grow into small planets. The Earth-like planets are large accumulations of solid particles that grow from the collisions of these planetesimals. In the outer regions of the disc, when a solid core becomes large enough (mass about 10 times the Earth-mass) it accretes gas and becomes a giant gas planet like Jupiter. The inner planets cannot accumulate gas as the temperature close to the proto-star is too high to allow gas accretion. The planet building phase is believed to take 10^7-10^8 years; the inner planets probably grow quickly, whereas the more distant gas giants require considerably longer. Current theories of planet formation are unable to explain the existence of Neptune in the solar system. A schematic of the formation of a planetary system is shown in Figure 7.9.

Planetary systems – always with intelligent life – around other stars have been in the realm of science fiction for a long time. But in 1995 extra-solar planets entered the realm of science. The discovery, in the late twentieth century, of large dusty discs around a number of stars spurred on the search for mature planetary systems. Detection of planets around other stars is a technically demanding problem. Planets are considerably smaller than stars and shine only by reflecting the light of their parent star. Suppose we were to look at the solar system from a planet orbiting the nearest star, Proxima Centauri, which is about 1.3 parsec or 4.3 light-years away. Jupiter would be the brightest planet but about 7000 times fainter than the Sun. The second brightest planet would be Venus, about 18000 times fainter than the Sun. Jupiter would appear about 3.5 arcseconds away from the Sun and this would be easily resolved with a small telescope on the planet. Venus however would be about 0.5 arcseconds from the Sun and it would be necessary to have a large telescope at a very good site to detect it. Fortunately it is not necessary to look at reflected starlight to detect planets, gravity lends a helping hand. A planet and a star orbit about their common centre of mass and the star 'wobbles' slightly under the gravitational influence of the orbiting planet. This wobble Doppler-shifts the wavelengths of the sharp absorption lines of chemical elements in the stellar spectrum. These shifts are small but for some stellar systems they are measurable with conventional spectrographs. This is what Michel Mayor and Didier Queloz of the University of Geneva did in 1995. They pointed their telescope with a spectrograph at the star 51 Pegasi and detected the small shift of 56 metres per second every 4.2 days in the apparent wavelength of the sharp absorption lines in the spectrum of this star. These observations suggested that there was a planet of at least half the mass of Jupiter orbiting 51 Pegasi. Since then, several research groups have joined the search and by mid-2000, 29 extra-solar planetary candidates had been discovered. A planet has also been discovered orbiting a pulsar. Most of these exoplanets are unusual when compared with the solar system planets. The minimum mass of these planets ranges from about half to a few times the mass of Jupiter. Moreover, these planets are very close to their parent star. This is at odds with the conventional theory of planet formation described above. It is possible that the orbits of these giant planets are decaying through (gravitational) tidal drag and the planets are spiralling towards the parent star.

In the closing months of 1999 a team of British astronomers made the first direct observation of the faint starlight reflected off a planet orbiting the star tau Boötis. This opens the interesting prospect of determining the principal constituents of the atmosphere of some exoplanet by detecting the lines formed by absorption of starlight by gases in the planetary atmosphere. Even more exciting discoveries lie ahead. Infrared telescopes planned to be launched in the first quarter of this century will be able to detect signatures of gases such as carbon dioxide, oxygen, ozone and water vapour in the atmospheres of planets in orbit around nearby stars. The large amounts of oxygen and ozone in the Earth's atmosphere were created by the blue-green algae in the primordial oceans of the Earth about 2.5 billion years ago. Detection of large quantities of these gases in the atmosphere of an exoplanet would be a strong indication of the presence of (algal) life on the planet with (maybe) intelligent life to follow.

8 Planck

Gravity is beautifully successful in describing the universe at macroscopic level but at microscopic level it is totally inadequate. By contrast, quantum theory is highly successful in describing matter on small scales. The detailed understanding of chemical reactions, lasers, microchips and nuclear weapons is based entirely on quantum physics. The concept of an atom originated in Greek antiquity, but it was suppressed under the baleful influence of Plato and Aristotle. In the early nineteenth century John Dalton resurrected 'atomism'. Dalton was a Quaker and was born in 1766 in Eaglesfield, Cumberland, in England. He recognised that atoms could help to understand the data being accumulated by chemical and physical experiments. Dalton's atoms were the smallest indivisible unit of a substance and they retained their chemical properties. He maintained that chemical reactions were just the rearrangement of these basic units of matter. Dalton presented his work in a two-volume thesis called *New System of Chemical Philosophy*, published between 1808 and 1827. This laid the basis of modern chemistry. Dalton's atoms were not accepted enthusiastically by all chemists of the early nineteenth century. However, it was soon realised that the chemists and physicists of the day had independently accumulated data which suggested the atomic nature of matter. In particular, physicists such as Maxwell and Boltzmann maintained that the pressure exerted by gas in a container could be explained if the gas was assumed to be a collection of hard spheres, like billiard balls, which bounced off the walls of a container in accordance with Newtonian mechanics. Pressure can be interpreted as the reaction of the walls to the force exerted by the bouncing gas atoms. Similarly, heat was the result of rapid random motion of atoms and molecules, and this motion could be described by a parameter called *temperature*. It is worth emphasising that this gas kinetic theory seeks to explain the *random* motion of gas atoms or molecules, unlike the theories of Newton and Einstein, which seek to explain the regular motion of bodies. In spite of the success of the gas kinetic theory, the very existence of atoms was disputed even in the last decade of the nineteenth century.

The quantum revolution was started by a very reluctant revolutionary. Max Karl Ernst Ludwig Planck (1856–1947) was born in Kiel, Germany, the sixth child of a distinguished jurist and professor of law at the University of Kiel. His family had a long tradition of devotion to the church, the state and scholarship. In 1867 the family moved to Munich and young Planck entered the city's renowned Maximilian Gymnasium where a teacher stimulated his interest in physics and mathematics. But at school Planck also excelled at other subjects. At 17 Planck had to choose a career and he decided to pursue a life of research in physics. Later he was to recall that the choice of physics was dictated by his early realisation that 'pure reasoning can enable man to gain an insight into the mechanisms of the world'. Planck was to conclude that the natural world was something absolute and independent of man, and the quest of laws which apply to this absolute world was the most sublime of pursuits. Planck entered the University of Munich in 1874 and after a spell at the University of Berlin, returned to Munich to receive his doctorate in 1879 (the year of Einstein's birth), at the unusually young age of 23. In 1892 he was appointed a full professor at the University of Berlin where he stayed the rest of his working life. Planck was soon recognised as one of the foremost thermodynamicists of his day.

One absolute of nature which impressed Planck deeply, even during his school days, was the law of conservation of energy sometimes called the First Law of Thermodynamics. Later during his university days he became equally convinced that the Second Law of Thermodynamics was also an absolute law of nature. The Second Law became the subject of Planck's doctoral thesis and lay at the core of the groundbreaking research that led him to the discovery of the quantum of action. In 1859–60 Gustav Robert Kirchhoff, an eminent scientist at the University of Berlin and Planck's future lecturer, had defined a black body as an object that re-emits all radiation incident upon it, that is, it is a perfect emitter and absorber of radiation. There was therefore something absolute about black body radiation. In the 1890s various experimental and theoretical attempts were being made to determine the spectral energy distribution of the black-body – the amount of energy emitted at different frequencies by a body at a given temperature. Planck was particularly impressed by the empirical formula found in 1896 by Wilhelm Wien and he made various attempts to derive the formula on the basis of the Second Law of Thermodynamics. According to the Wien Law, the energy emitted in any waveband is proportional to the frequency (or wavelength) of the emitted radiation. Thus the theory predicted that the emitted energy should increase as the frequency increased or as one went towards the blue end of the spectrum, reaching infinite emission at high frequencies. This absurd consequence was dubbed the 'ultraviolet catastrophe'. Observations of hot bodies suggest that although the energy emission initially increases with frequency of the emitted radiation, at high frequencies the emitted energy is actually lower, with a peak somewhere in between (see Figure 7.4). The position (or the frequency) of this turnover depends on the temperature of the radiating body.

Planck arrived at his theory of black-body emission after a suggestion by Ludwig Boltzmann that he consider a statistical approach to the problem. Maxwell, after all, had given a statistical interpretation of the Second Law of Thermodynamics. Initially Boltzmann had fiercely resisted this interpretation but he eventually came to accept it. The statistical interpretation is based on the concept of discrete atoms and molecules. Boltzmann believed that there was 'no reason why energy shouldn't also be regarded as divided atomically'. Boltzmann was violently opposed by other theoretical physicists and in despair he killed himself before his ideas were widely accepted. Planck, a physicist of the old school, was opposed to atomic theory and he initially rejected Boltzmann's suggestion. However, he eventually relented, accepted Boltzmann's suggestion and produced a beautiful theory of energy distribution of black-body radiation, which was presented to the German Physical Society on 19 October 1900.

To formulate his theory of black-body emission Planck had to abandon one of his most cherished beliefs, that the Second Law of Thermodynamics was an absolute law of nature. Instead he had to accept, as Maxwell had shown, that the Second Law was a statistical law. He also had to accept (as Boltzmann had suggested) that electromagnetic radiation is delivered in packets called *quanta*. The mathematical relationship lying at the heart of this work is

$$\varepsilon = h\nu$$

where ε is the energy of a quantum, ν is the frequency of radiation and h is a universal constant of action, now known as *Planck's constant* ($h = 6.6262 \times 10^{-34}$ joule second). The physics community of the early twentieth century reacted with intense scepticism to Planck's new idea. Light and heat chopped into energy packets and acting like particles was just preposterous. Five years later, in 1905, Einstein (still an obscure physicist) used Planck's ideas to develop the theory of the photoelectric effect. It was years before Planck's achievement was generally recognised but physicists increasingly realised that because Planck's constant, recognised as the quantum of action, was not zero but had a finite value, the microscopic world (the world of atomic dimensions) could not be described by classical mechanics. A profound revolution in physics was taking place. Planck was awarded the Nobel Prize in physics in 1918 for his quantum theory.

Wars and the intolerance of Nazi Germany blighted Planck's life. His elder son was killed in action in the First World War. He attempted to persuade Hitler to reverse his obnoxious racial polices. In this he failed but he chose to remain in Germany during the Nazi period and was fortunately spared the penalties for this act of bravery. His younger son however was not so fortunate; he was implicated in the attempt to assassinate Hitler on 20 July 1944 and in early 1945 he died at the hands of the Gestapo. After the war Planck and his family were moved, by the Allied Army Command, to Göttingen where he died in 1947.

These days almost all of physics is described by the theory of quantum mechanics, the exception being gravitation. However, there are regimes that are both exceptionally small (requiring quantum mechanical description) and exceptionally massive (requiring general relativistic mechanism); for example, the singularity in a black hole and the beginning of the universe. Dimensional analysis can be used to identify the regimes in which general relativity may break down and a quantum interpretation of gravity (or quantum gravity) may be required. The regimes must involve the three fundamental constants of nature, G (the Newtonian constant of gravitation), h (Planck's constant) and c (the speed of light). The length scale of the regime at which the two theories have to be considered together is given by

$$hG/c^3 \approx 10^{-35} \text{ metres}$$

which is called the *Planck length*.

The laws of quantum gravity are not known at present and the search is difficult because there are no experimental guideposts nor has it been possible to recognise any properties of the universe that may give clues to quantum gravity, assuming such clues exist. Lacking definite clues, speculations abound – gravity is about space-time, and quantum theory involves fluctuations, therefore quantum gravity will involve fluctuations of space-time, called the probabilistic froth. At a distance of Planck length from a singularity, like that in a black hole or the big bang, space-time is believed to dissolve into this random probabilistic froth. Close to and inside the singularity all curvatures of space-time are allowed, and the laws of quantum gravity should give the probability of these curvatures and describe the nature of this froth. The laws of quantum gravity may also give the probability of a singularity resolving into classical (nonquantum) regions of space-time – that is, give birth to a universe.

Unlike the general theory of relativity, quantum mechanics deals with chance and probability. The German physicist Werner Heisenberg identified the probabilistic nature of quantum mechanics in 1927 – he called it the *uncertainty principle*. In a quantum world, particles do not possess independent properties like momentum and position; they carry a mixture of the two and the two can *never* be completely unravelled. Heisenberg showed that this uncertainty was a fundamental feature of quantum theory. This places previously unsuspected restrictions on the motion of atoms and sub-atomic particles. Only those states of motion are possible in which action is a whole-number (non-zero) multiple of Planck's constant. The uncertainty principle asserts that the momentum and position of a particle cannot be simultaneously determined unambiguously. This is a fundamental (quantum) property of a particle and not just an expression of limitations of experimental techniques. In the general theory of relativity space-time is both smooth and continuous and the equivalence principle asserts that the world-line of an object is defined *exactly*. In other words, the momentum and position of the object are defined exactly. This principle is the cornerstone of general relativity. But this requirement is inconsistent with the Heisenberg uncertainty principle, which is the cornerstone of quantum mechanics. Thus the theories of general relativity and quantum mechanics are incompatible at a fundamental level. At its basic level the universe appears to obey the laws of quantum mechanics (for reasons not understood at present) and it seems likely that general relativity may not contain the last word on the nature of gravity. However, a quantum theory of gravity must be able to make predictions, which should asymptotically approach those of general relativity.

Einstein with Planck and Erwin Schrödinger (the originator of wave mechanics) remained opposed to the indeterministic or statistical world-view introduced into physics by the advent of quantum mechanics. But this was not a petulant unwillingness or inability to accept the experimental reality unfolding in the 1920s and 1930s – Einstein and Schrödinger continued to make fundamental discoveries in quantum mechanics throughout their working life. Their opposition was based on conflict with their deeply held belief that the physical universe was an objective entity existing independently of man; the observer and the observed are not intimately linked as required by quantum mechanics. Einstein (like Niels Bohr, one of the pioneers of quantum theory) was also concerned that the fundamental principle(s) pinning together the quantum theory had not been identified.

We now know that there are four fundamental forces of nature: gravity, electromagnetism, weak nuclear force and strong nuclear force. Gravity is an attractive force, which holds the planets in their orbits, prevents the stars from exploding and holds galaxies and clusters of galaxies together. The electromagnetic force holds the atom together. It determines the orbits of electrons in atoms and effectively governs the laws of chemistry. Within the nucleus of an atom the strong and weak (nuclear) forces dominate. The strong force binds together protons and neutrons, counteracting the repulsive electromagnetic force between protons. There is a delicate balance between the attractive strong force and the repulsive electromagnetic force. When the number of protons in a nucleus is greater than about 100, the strong nuclear force cannot hold the electromagnetic repulsion and the nucleus disintegrates. This is why there are only about 100 naturally occurring elements. The strong force keeps the stars shining and can cause catastrophic destruction when released in a nuclear explosion. The weak force is perhaps the most enigmatic. This force determines the rate at which hydrogen is burned to helium in stars and the Sun. The radioactive decay of otherwise stable particles is caused by the weak force, and the intense heat in the interior of the Earth is partially due to the weak force causing radioactive decay in the core and mantle of the Earth. The destructive nature of this force is displayed in volcanic eruptions. In some sense the existence of life on Earth is possible because of (the weakness of) the weak force.

Why are there four forces of Nature? In the last decade of the nineteenth century it was realised that electricity and magnetism were manifestations of the same force. Einstein was convinced that there should be a single description of his geometric theory of gravity and Maxwell's theory of light. It is not surprising that Einstein should have chosen to unite gravity and electromagnetism, as these two forces were subjects of intense investigation during his lifetime. We now know that Einstein was on the wrong track, but his belief that the forces of nature ultimately must be united by a single physical principle has had a strong hold on scientific thinking from the latter half of the twentieth century. An early attempt to unite quantum mechanics and special relativity was made in the 1930s. It was realised that when the velocities of colliding particles approach the speed of light, quantum mechanics 'breaks down' in the sense that it predicted a useless series of infinities. A way round this was found by Richard Feynman and Julian Schwinger, two American physicists, and the Japanese physicist Shinichiro Tomonaga. They developed the theory of Quantum ElectroDynamics (QED). In this theory the charge and mass of the electron are *redefined* to cancel the troublesome infinities, by exploiting a special property of Maxwell's equations called 'gauge symmetry'[1]. The conservation or invariance of electric charge follows naturally from the gauge symmetry of the equations. In the QED theory a collision between two electrons is explained as an exchange of a photon between the colliding particles (unlike gravitational interaction, which is described in terms of distortion of space-time caused by mass or energy of a body). The theory only works for electrons and photons but it predicts results that agree with observations with amazing precision. Feynman, Schwinger and Tomonaga shared a Nobel Prize for physics in 1965 for their discovery of QED.

The description of quantum processes in terms of exchange of particles is now basic to quantum theory. The force 'field' around a particle is described in terms of continual absorption and emission of fundamental particles, known as field quanta. The field quantum of the electromagnetic field is the *photon* – a 'particle' of light. The success of QED encouraged physicists in the late 1960s to apply the reasoning behind QED to the weak interaction. They conjectured that the weak force might also be caused by exchange of a new set of particles, called

W-particles (W for weak). However, this theory had no gauge symmetry and it was plagued by troublesome infinities, not unlike the infinities that had plagued the early attempts to formulate a theory of electromagnetic collision. However, the American physicist Stephen Weinberg and the Pakistani physicist Abdus Salam, working independently and applying a concept developed earlier by Sheldon Glashow, reasoned that at high enough energy or temperature – such as that which occurred soon after the big bang – the electromagnetic and weak fields would be indistinguishable; this 'combined' field is called the electroweak field. As the temperature drops the electromagnetic and the weak field freeze out of the electroweak field. The reverse process, the unification of the electromagnetic field and the weak field, occurs in high-energy collisions of elementary particles. This process of unification has now been studied in detail at the Large Electron Positron collider at CERN, the European particle physics laboratory near Geneva in Switzerland. The freezing-out of individual fields from a combined field is called symmetry breaking. The electroweak theory uses a very sophisticated form of gauge symmetry, which possesses far more symmetries than Maxwell's theory. This theory treats electrons and neutrinos as one 'family'; they are considered two sides of the same coin. But it does not explain why there are three redundant electron families (electron, muon and tau, all similar except in their mass and their associated neutrinos). In 1979 Weinberg, Salam and Glashow were awarded the Nobel Prize for physics for this first unambiguous step towards unification of the four forces of nature.

The success of QED and electroweak theories encouraged physicists to apply gauge symmetry to the strong force. In the 1950s and 1960s, physicists had discovered hundreds of strongly interacting particles (called hadrons) in their particle colliders. Order began to emerge in the early 1960s when Murray Gell-Mann (American) and Yuval Neéman (Israeli) showed that these hadrons occurred in patterns of eight (Gell-Mann called this the Eightfold Way – the name of the Buddhist path to wisdom). Later Gell-Mann and George Zweig showed that the Eightfold Way arises naturally if a combination of triplets of subnuclear particles, dubbed 'quarks', formed the hundreds of hadrons found in the laboratories. The quarks are held together by the quantum field particles called gluons. The force holding the quarks together is called the 'colour' force and the resulting theory is called 'Quantum ChromoDynamics' or QCD – the theory of 'colour' interactions. Gell-Mann was awarded the Nobel Prize for physics in 1969 for his work on QCD. The quark theory was developed for the next 25 years and gained enough respectability to be called the 'Standard Model'. The discovery of top quark, in 1995, was the last piece of the jigsaw puzzle to complete this Standard Model. Unfortunately the Standard Model is defined by 19 parameters and these cannot be predicted theoretically: they have to be inserted *ad hoc* into the model and their values have to be adjusted from the measured properties of elementary particles. The Standard Model thus lacks the elegance of general relativity and it is now considered to be only an intermediate step towards a true theory unifying the electromagnetic, weak and strong force.

In 1974 two American physicists, Sheldon Glashow and Howard Georgi, proposed that just as the electromagnetic field and the weak field combine at high temperature into the electroweak field, at even higher temperature the three nongravitational fields – electromagnetic, weak and the strong fields – would unify. Glashow, Georgi, Helen Quinn and Stephen Weinberg developed this proposal into the 'grand unified theory' (GUT), a theory to link the electrons, the neutrinos and the quarks and the corresponding field quanta of electromagnetic, weak and strong force, namely the photon, W-particle and gluons. But GUT is difficult to test at present because the energy at which the strong force and the electroweak force are united is beyond the reach of the present generation of particle colliders. GUT makes only one prediction that can be tested at present – it predicts that the proton (made of three quarks) has a finite lifetime and will eventually decay into electrons and other particles. Several groups of physicists in different countries are looking for evidence of proton decay with detectors down deep mineshafts. After several years of operation no conclusive proof of proton decay has been found yet. These experiments suggest that the mean life of the proton is longer than 10^{32} years. GUT is also littered with scores of arbitrary and undetermined parameters (the masses of quarks and leptons, for instance). The current consensus of opinion is that GUT is also not the ultimate unified theory.

An interesting consequence of this deepening understanding of the nature of quantum forces has been the emergence of our understanding of the 'origin of mass'. The theories of gravitation of Newton and Einstein do not provide a consistent explanation of the mass of particles. In order to give particles mass, quantum 'mechanists' postulate a universal background field, called the Higgs field (after Peter Higgs, a physicist at the University of Edinburgh). When a particle moves through this field the field becomes locally distorted, and this distortion generates the particle's mass. The field quantum of the Higgs field is the Higgs particle, which is the clustering of the Higgs field without the presence of a particle. The Higgs particle (and therefore the evidence for the Higgs field) has not been discovered yet. The Standard Model cannot predict the mass of the Higgs particle with any certainty so it is difficult to identify the energy regime in which the search should be conducted. In late summer 2000 tantalising evidence for the Higgs particle (Higgs boson) was seen at the Delphi detector of LEP. But corroborating evidence at other detectors was not present and further observations did not confirm the detection. Clearly this is a task for the next generation of particle colliders. The total mass of all particles is not made up of just the distortion of the Higgs field; for example, the bulk of the mass of particles such as protons and neutrons comes from the energy which holds together the constituent quarks of these particles. The Higgs field theory of the origins of mass of particles unfortunately does not explain why the elementary particles have the masses that are observed. Also, the theory raises a further question: what 'causes' the Higgs field?

The latest theory in the running for the unified field theory is the superstring theory. String theory was first formulated in 1970, by Yoichito Nambu of the University of Chicago. He proposed that the point particles, the basic building blocks of matter, should be replaced by tiny, one-dimensional filaments called strings. The theory proposed that these strings are the ultramicroscopic ingredients of the elementary particles. This proposal was made to make sense of the hundreds of hadrons being discovered in various collider laboratories. In this theory the strings interact by breaking and reforming in a well defined way and the plethora of 'elementary' particles are simply different vibration modes (resonances) of the same string, with no resonance any more elementary than another. At the length scale probed by current experiments the strings would appear point-like, or like particles. The formulation of the string theory required 26 dimensions. The physicists of the late 1960s had come to terms with the four dimensions of general relativity but the additional 22 dimensions were just too many for the 1970s. The theory also predicted unusual particles, particles that were unexpected and 'undesirable', like tachyons that could only move faster than the speed of light. This was 'Star Trek' stuff, good for television, bad for physics. By the mid-1970s, support and enthusiasm for the string theory had begun to wane and for next decade the young and productive physicists turned to the rapidly developing electroweak and GUT theories.

There were two exceptions, Michael Green of Queen Mary College in London (England) and John Schwarz of the California Institute of Technology (USA). In 1976 they proposed that the string theory was more than a theory to explain the strong interactions, as originally conceived by Nambu: it was actually a theory of the universe! The unexpected and undesirable particles were real particles. The new development was the introduction of the concept of 'supersymmetry' into string theory, making it a superstring theory and possessing all possible symmetries. Superstring theory employs the Kaluza–Klein concept to unify gravitation and the gauge interactions, by using higher dimensions. In this theory the field quantum of the gravitational field is a massless particle called the graviton. Its long-wavelength interactions are similar to those described by Einstein's theory of general relativity. It is tempting to suggest that general relativity is a natural consequence of superstring theory. In this theory the embarrassing tachyons disappear. Superstring theory was received with extreme scepticism. But in early 1980s the Kaluza–Klein theory became fashionable and physicists gradually overcame their prejudice against higher dimensions. According to superstring theory the everyday world has 9 + 1 space-time dimensions, six of which have curled up and are small and compact, but time and the three spatial dimensions have expanded to infinity.

By the early 1990s, superstring theory had moved from an interesting and beautiful curiosity to a dominant position in the world of physics. The Standard Model does not attempt to provide a fundamental description of physics at the shortest distance scale. String theory attempts just this short distance scale description. In string theory, fundamental particles are no longer described as points but arise as different modes of excitation of an extended string-like object. This simple paradigm has two immediate consequences: the elementary particles are unified (they are different modes of excitations) and at distances large compared with the Planck length the string theory describes the familiar laws of general relativity. *But the fact that a string has a nonzero size leads to a new physics at smaller scale.* It appears that supersymmetry is a crucial ingredient in the string theory. The concept of symmetry was originally developed to unify matter and radiation into a single theory. It is increasingly being realised that symmetry is essential to construction of physical laws; recall that Noether discovered the three 'Newtonian conservation laws' from the symmetry of Newton's laws of motion. Nature actually seems to demand symmetry in the laws of physics. The beauty of Einstein's four-dimensional general theory is in its symmetry between space and time dimensions: space can be rotated into time. Supersymmetry is not only aesthetically pleasing, it has become an essential ingredient in almost any attempt to unify the four forces of nature. The superstring theory has so many symmetries that it can include all the symmetries of the electroweak and GUT theories. A variant of the superstring theory, called M-theory (M for membrane), is a supersymmetric theory described by 11 dimensions – 10 spatial dimensions and 1 time dimension. At low energies (the everyday energies) seven spatial dimensions are curled up to infinitesimally small size, and time and three spatial dimensions are stretched out to infinity. The enthusiasm for this theory as the unified theory is based on an almost firm conviction that all the known symmetries of the universe, and those yet to be discovered, are contained in this theory. In this theory the masses of particles or the field quanta of forces are determined by the vibration patterns of the curled-up space dimensions. At present there is no experimental evidence for (super)string theory but this may be because the energies attained in the present generation of particle colliders are too low. The Large Hadron Collider (LHC) being built at CERN in Switzerland (and expected to be operational in 2005) will accelerate two beams of protons in opposite directions, so that when the protons in the beam collide energies of 14×10^{12} electron volts will be released. This energy is equivalent to a temperature of about 10^{17} degrees. It is believed that this temperature was reached at about 10^{-12} seconds after the big bang (see Figure 7.3). In the debris of these collisions, physicists hope to find new particles, particles that would correspond to the low-energy vibrations of the superstring.

For higher-energy vibrations it will be necessary to look at the universe at its very beginnings, because machines to produce these higher energies are beyond the capability (and imagination) of present technology. The superstring theory views the big bang as a by-product of a much more violent explosion, the breakdown of an 11-dimensional universe into a four-dimensional universe. This, of course, begs the question, what caused the 11-dimensional universe to explode? That question can't be answered at present. At the end of this primordial explosion seven dimensions curled up and four expanded to infinity, and the big-bang evolution proceeded as described in the previous chapter. The manner in which the seven dimensions rolled up is significant, for it determined the values of the constants of nature and the masses of the elementary particles, but this is not fully understood yet.

There is thus a very close link between astrophysics/cosmology and elementary particle physics. A significant example of this vital link is provided by the connection between helium abundance and the number of types of neutrinos. The big-bang theory accounts very satisfactorily for the synthesis of helium in the universe. The synthesising process depends on the number of types of neutrinos. The observed mass fraction of about 25% of helium suggests that the types of neutrinos cannot exceed four and the most probable number is three. The Standard Model suggests that each neutrino is associated with a family of elementary particles. The model does not predict the number of families and in principle it is possible to find an almost infinite number of families as higher energies are explored. But astrophysics coupled with cosmological arguments states with certainty that there can, at most, be four families and three are most probable. This is a radical statement: it means that astronomical observations can predict something as fundamental as the number of types of elementary particles.

The astronomical prediction of number of families of elementary particles was made in the mid-1970s. By the late 1980s it was possible to test this in the laboratory. The tests were done by studying the decay of the Z^0 particle. This particle can decay into any number of neutrino species and the number of species is determined by the lifetime of Z^0. The detectors at the Large Electron Positron (LEP) collider at CERN in Geneva have measured this lifetime very accurately and the results suggest that there are 2.99 ± 0.02 neutrino families. This is in excellent agreement with the cosmological theory of helium synthesis. Cosmology and particle physics are thus intimately linked, and predictions in one area can be verified by experiments in the other. The PLANCK and the MAP experiments to map the anisotropy of the microwave background radiation – the afterglow of the big bang – will provide one of the very few ways of testing physics at ultra-high energies. These satellite missions are thus of crucial importance in the development of fundamental theories of physics.

String theory suggests a link between the Newtonian gravitational constant G and the constants of physics obtained from quantum theory (e.g. Planck's constant). At present it is the only theory which makes this link. The predicted value of G is about 100 times larger than the observed value, but the point is that for the first time a possible link has been predicted between gravitation and quantum theory. New developments also suggested that the 'curled-up dimensions' of the superstring theory may have a length scale of the order of 0.1 millimetre without contradicting the existing experimental results. But this suggests that the inverse square law of gravitational interaction could be dramatically modified below some distance smaller than 0.1 millimetre. Also a generic prediction of various string theories is that there may be other gravitational-strength interactions than the one described by Einstein's general relativity. These new interactions would violate the universal free fall. The most sensitive probe of these new forces is the tests of the equivalence principle. The current accuracy of 10^{-12} of these tests does not diminish the possibility of violations at lower levels – indeed theoretical speculations suggest violations at levels lower than 10^{-12}.

The superstring theory is a powerful conceptual construct but it is far from being an experimentally verified theory. Nonetheless, there is here the potential to resolve the incompatibility between quantum mechanics and general relativity. The motivations for finding this theory are very strong. Besides the aesthetic satisfaction of achieving consistency between quantum mechanics and general relativity, there is the enormous reward of understanding the origins of a universe, or even the Universe.

The superstring theory has been called the 'Theory of Everything' – a theory that may provide a unified explanation of everything, from the smallest particle of matter to the largest galaxy. This has led to (extravagant) claims that 'there are grounds for cautious optimism that we may now be near the end of search for the ultimate laws of nature'[2]. Similar sentiments were being expressed in the early days of the twentieth century. In 1903 Michelson commented: 'The more important fundamental laws and facts of physical science have all been discovered, and these are now so firmly established that the possibility of their ever being supplemented in consequence of new discoveries is exceedingly remote.' Within a decade the 2000-year-old deterministic physics and the certainty of absolute space and time had been swept aside.

The twentieth century has been a period of extraordinary increase in our understanding of laws of nature and new discoveries are being made every day. It is possible that in the next decade or two the constituents of the dark matter will be identified, the mechanism responsible for the asymmetry between matter and antimatter will be recognised and the formation of galaxies will be understood, but a number of formidable questions have to be answered before writing 'finis'. For example, it is not known that superstring theory will be able to deal with the singularity that occurs in a black hole. These singularities are a natural consequence of general relativity or the 'seeds of its own destruction'[3]. There is also an embarrassing lack of understanding of the 'origins' of the laws of quantum mechanics. Quantum theory is not a theory of principles; the laws of quantum mechanics are purely phenomenological. They can predict the results of quantum processes with exceptional precision, but it is not known where these laws come from or why they work so well. Similarly, the principles stitching together the discoveries made in string theory have not been established. Also, where have space and time come from? This is the sort of question children ask but adults have not been able to answer. While it is encouraging to feel that Einstein's 'final theory' is in sight it is more likely that we are on the brink of another major revolution in physics – a revolution as profound as that which occurred at the beginning of the twentieth century.

Chronology

Milestones in the evolution of the laws of motion and gravitation

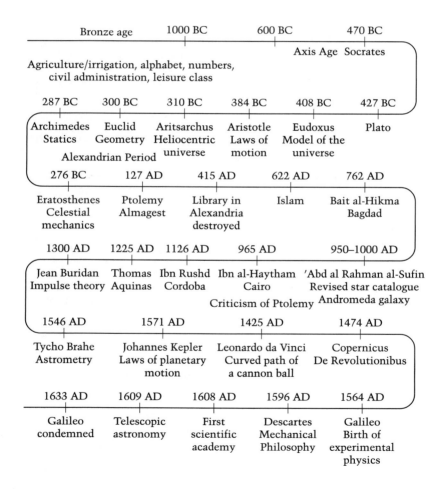

CHRONOLOGY

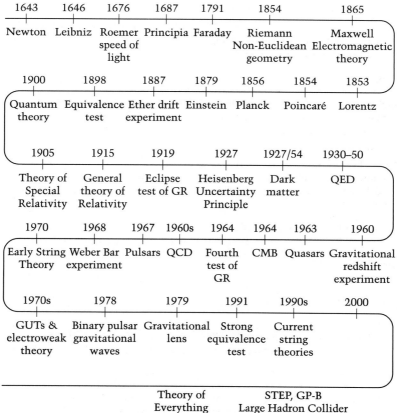

1643	1646	1676	1687	1791	1854	1865
Newton	Leibniz	Roemer speed of light	Principia	Faraday	Riemann Non-Euclidean geometry	Maxwell Electromagnetic theory

1900	1898	1887	1879	1856	1854	1853
Quantum theory	Equivalence test	Ether drift experiment	Einstein	Planck	Poincaré	Lorentz

1905	1915	1919	1927	1927/54	1930–50
Theory of Special Relativity	General theory of Relativity	Eclipse test of GR	Heisenberg Uncertainty Principle	Dark matter	QED

1970	1968	1967	1960s	1964	1964	1963	1960
Early String Theory	Weber Bar experiment	Pulsars	QCD	Fourth test of GR	CMB	Quasars	Gravitational redshift experiment

1970s	1978	1979	1991	1990s	2000
GUTs & electroweak theory	Binary pulsar gravitational waves	Gravitational lens	Strong equivalence test	Current string theories	

Theory of Everything

STEP, GP-B
Large Hadron Collider
LIGO, VIRGO,
GEO-600, TEMA
LISA

Notes

CHAPTER 1 **Aristotle**
1 The Babylonians were perhaps the earliest systematic stargazers. Astronomical records of times of moonrise, date of the new moon and the dates of heliacal rising and culmination of a number of stars, evenly spaced in the sky, were compiled from about 1100 BC.
2 The normal relative motion of planets is towards the east. Occasionally planets appear to reverse their course, drifting *backwards* (westwards) against the background of stars. This retrograde motion can last for several months.

CHAPTER 2 **Kepler**
1 Quoted from *Nicholas Copernicus on the Revolutions*, Translation and Commentary by Edward Rosen, The John Hopkins University Press, 1992.
2 In India the notion of fixed Earth was challenged in 500 AD by Aryabhata (b. 476 AD). In his book on astronomy, the *Aryabhateeya*, there is an explicit mention of Earth's rotation about its axis. But the idea of the 'fixed Earth' was firmly established in India and Aryabhata was ignored.
3 Quoted from *Theoretical Concepts in Physics*, by M.S. Longair, Cambridge University Press.

CHAPTER 3 **Galileo**
1 Quoted from *The World within the Worlds* by J.D. Barrow, Oxford University Press, 1988.
2 This story may have been perpetuated by Vincenzio Viviani, Galileo's biographer. Swinging lamps were installed in the cathedral of Pisa in 1587, five years after Galileo is supposed to have timed their swing!
3 Quoted from *Great Experiments in Physics* (ed. M.H. Shamos), Dover Publications, Inc., 1959.
4 Modern research suggests that the true father of the telescope may be Leonardo da Vinci. He certainly suggested that a special optical glass would be required to examine the moon's surface.

CHAPTER 4 **Newton**

1. Quoted from *Newton* by I.B. Cohen and R.S. Westfall, A Norton Critical Edition, 1995.
2. *Matter and Motion* by James Clerk Maxwell, Routledge/Thoemmes Press, 1996.
3. Newton calculated that the world was created on 3998 BC. This should be compared with the date of 4004 BC obtained by Archbishop James Ussher (a contemporary of Newton).

CHAPTER 5 **Einstein**

1. Based on a scenario in *'The Lighter Side of Gravity'* by Jayant V. Narlikar, Cambridge University Press, 1996.
2. 'The relative motion of the Earth and the luminiferous ether' by Albert A. Michelson, Master, U.S. Navy. *American Journal of Science* **22**, 120, 1881.
3. Quoted from *Einstein Lived Here*, by Abraham Pais, Oxford University Press, 1994.
4. In his 1905 paper *On the electrodynamics of moving bodies* Einstein states that one of the rationales for his paper was 'the unsuccessful attempts to discover any motion of the earth relative to the "light medium,"'. This suggests that Einstein may not have been aware of the Michelson–Morley experiment but he was certainly aware of its implications. Quoted from *The Principle of Relativity* by H. A. Lorentz, A. Einstein, H. Minkowski and H. Weyl, transl. W. Perrett & G.B. Jeffery, Methuen, 1923.
5. To obtain the total energy of a particle it is necessary to take the square root of each side of this equation to obtain $E = \pm\sqrt{(c^2p^2 + m_0^2c^4)}$. In 1928 Paul Dirac (1902–1984) recognised the importance of the negative square root and developed the theory of antimatter. The first antiparticle (positron – a positively charged electron) was discovered in 1933 by the American physicist Carl Anderson. In 1936 he was awarded the Nobel Prize for physics.
6. Recent discovery of previously unnoticed proofs of Hilbert's papers cast a different light on the question of priority regarding the general theory. It appears that between the submission of his paper on 20 November and its publication Hilbert made significant changes in his paper and these changes may have been influenced by the progress Einstein had made after his visit to Göttingen. See Leo Corry, Jürgen Renn and John Stachel, 'Belated decision in the Hilbert–Einstein priority dispute' in *Science* **278**, 14 November 1997.
7. The name 'black hole' was introduced by the American physicist John Archibald Wheeler in 1968 in a lecture to the American Association for the Advancement of Science. Unfortunately the name has caused some confusion; Wheeler meant a potential well or a large gravitational attraction from which even light cannot escape, not a physical hole into which things can fall.
8. See 'Gravitation theory' by Clifford M. Will in *Scientific American* (November 1974) for a nontechnical description of alternative theories of gravitation.

9 William Clifford, 'On the space theory of matter', *Proceedings of the Cambridge Philosophical Society* **2**, 157–158, 1876.
10 Quoted from *Black Holes and Time Warps: Einstein's Outrageous Legacy* by Kip S. Thorne, Papermac, 1995.

CHAPTER 6 Dicke
1 Quoted from F.W. Dyson, A.S. Eddington, and C. Davidson, *Phil. Trans. R. Soc. A*, **220**A, 291–333, 1920.
2 Quoted from *The Hipparcos and Tycho Catalogue* ESA SP-1200, Volume **1**, June 1997.
3 Quoted from Clifford M. Will, 'The confrontation between general relativity and experiment: an update', *Physics Reports* **113**, No. 6, 1984.

CHAPTER 7 Hubble & Eddington
1 It is interesting to note that a similar measurement was undertaken in China in 725 AD by the Buddhist scholar monk I-Hsing and an official astronomer Nankung Yüeh. The aim was to fix a terrestrial length measurement (the *li*) in terms of astronomical units. Towards the end of the seventeenth century a Christian missionary, Antoine Thomas (1644–1709) repeated the measurement, at the suggestion of Khang-Hsi Emperor (P. Beer *et al.*, *Vistas in Astronomy* vol. **4**, 3, 1961). This was about 90 years before the French Academy undertook similar measurement.
2 Astronomical distances are expressed in Astronomical Units (AU) (the average distance from the Earth to the Sun) and the parsec (pc). 1 AU = 1.5×10^{11} metres and 1 pc = 3.1×10^{16} metres. A light-year is also often used to express distance: 1 light-year = 9.5×10^{15} metres.
3 Fred Hoyle coined the name 'hot big bang' during a British Broadcasting Corporation (BBC) radio broadcast in 1950. Hoyle used this name in a rather pejorative manner to describe the Lemaître–Einstein evolutionary model, as opposed to his own steady-state model. In 1995 the astronomy magazine *Sky & Telescope* organised a competition to find an alternative for the 'big bang', because in the USA big bang was not considered to be 'politically correct'. About 10 000 entries were received, but the judges decided to stay with the big bang!
4 1 solar mass = 1.989×10^{30} kilograms.
5 Quoted from J. Michell, *Philosophical Transactions of the Royal Society*, **74**, 35–57, 1784.

CHAPTER 8 Planck
1 An equation is defined at every point in space and time. If the equation remains unchanged for the *same* rotation at every point in space and time then the equation is said to have global symmetry. But if the equation remains

unchanged for *different* rotation at every point then it is said to have gauge symmetry.
2 Stephen Hawking in *A Brief History of Time, from the Big Bang to Black Holes*, Bantam Press, 1988.
3 Dennis Sciama, quoted in J.D. Barrow, *The World Within the World*, Oxford University Press, 1988.

Suggestions for further reading

Barrow, J.D. 1988. *The World Within the Worlds*. UK. Oxford University Press

Begelman, B. & Rees, M. 1998. *Gravity's Fatal Attraction: Black Holes in the Universe*. USA. Scientific American Library, W.H. Freeman & Co.

Chandrasekhar, S. 1996. *Newton's Principia for the Common Reader*. UK. Oxford University Press

Cohen, I.B. & Westfall, R.S. 1995. *Newton*. USA. W.W. Norton & Co., Inc.

Coveney, P. & Highfield, R. 1991. *The Arrow of Time*. UK. HarperCollins Publishers

Drake, S. 1978. *Galileo at Work – His Scientific Biography*. USA. Dover Publications, Inc.

Feynman, R.P. 1965. *The character of physical law*. UK. Penguin Group

Goodstein, D.L. & Goodstein, J.R. 1966. *Feynman's Lost Lecture*. USA: The Courier Company, Inc.

Greene, B. 1999. *The Elegant Universe*. UK. Jonathan Cape

Joseph, G.G. 1994. *The Crest of the Peacock, non-European Roots of Mathematics*. England: Penguin Books Ltd

Kaku, M., & Thompson, J. 1997. *Beyond Einstein*. UK. Oxford University Press

Kragh, H. 1996. *Cosmology and Controversy: The Historical Development of Two Theories of the Universe*. USA. Princeton University Press

Lorentz, H.A., Einstein, A., Minkowski, H., & Weyl, H. 1923. *The Principle of Relativity*, UK. Methuen & Co. (transl. Perrett, W. & Jeffery, G.B.)

Motz, L. & Weaver, J.H. 1988. *The Concepts of Science. From Newton to Einstein*. USA. Plenum Press

Narlikar, J.V. 1996. *The Lighter Side of Gravity*. UK. Cambridge University Press

Pais, A. 1982. *Subtle is the Lord ... The Science and the life of Albert Einstein*. UK. Oxford University Press

Ridley, B.K. 1994. *Time, Space and Things*. UK. Cambridge University Press

Schutz, B.F. & Will, C.M. 1993. *Gravitation and General Relativity*. Encyclopedia of Applied Physics, ed. G. L. Trigg, Vol. 7, pp. 303. USA: VCH Publishers, Inc.

Shamos, M.H. (ed.) 1959. *Great Experiments in Physics*. USA. Dover Publications, Inc.

Singer, C. 1959. *A Short History of Scientific Ideas*. UK. Oxford University Press

Smoot, G & Davidson, K. 1993. *Wrinkles in Time*. UK. Little, Brown & Co.

Thorne, K.S. 1994. *Black Holes and Time Warps*. USA. W.W. Norton & Co.

White, M. 1997. *Isaac Newton, the Last Sorcerer*. UK. Fourth Estate Ltd

Will, C.M. 1974. *Gravitation Theory*. USA. Scientific American (November 1974)

Will, C.M. 1993. *Was Einstein right?* UK. Oxford University Press

Wilson, R. 1997. *Astronomy through the Ages*. UK. Taylor & Francis

Index

3C273, 214, 279
3C279, 214
3C48, 279

51 Pegasi, 322

'Abd ar-Rahman al-Sufin, 20, 260
Abu al Barakat al-Baghdadi, 51
Academia del Cimento, 78, 79
Académie des Sciences, 79, 84, 124, 248
Accademia dei Lincei, 70, 77
accelerated motion, 63, 64, 155, 156
accretion, 281, 295, 310, 319, 321
accretion disc, 281, 282, 285, 296
action at a distance, 85, 88, 120, 132, 184
active galaxies, 172, 235, 283
Adams, John Couch, 123
Adams, Walter, 269
Adelard, 26
Adelberger, Eric, 230
Adolf, Gustaf, 36
age of Earth, 262
age of the universe, 196, 258
Airy, 123
al-Baghdadi, Abu al Bareket, 51
al-Battani, Mahammad, 21, 28, 35, 37
al-Ma'mun, 19
Almagest, 15, 20, 27, 28
al-Mansur, 19
Alpher, Ralph, 201, 265, 270
Ampère, André, 134
analytical geometry, 56
Anaxagoras, 7
Andromeda galaxy, 20, 260, 270, 284, 290
angular momentum, 46, 106, 107, 277, 296, 318, 320

Antares, 302
Apollonios, 49
Aquinas, St Thomas, 25, 26
Archimedes, 13, 14, 15
Aristarchus, 29, 30, 48
Aristotle, 2, 5, 15, 21, 51, 64
asteroids, 124, 200, 247, 321
astrology, 19, 22, 49
astrometry, 165, 212
Astronomical Unit, 124, 346
atom, 7, 134, 325
Augustine, Saint, 26, 266
Averroes, 21
Avicenna, 51

Baade, Walter, 278
Bait al-Hikma, 19
Barnothy, Jeno, 181
Bell, Jocelyn, 308
Bellarmino, Roberto Cardinal, 71
Benedetti, Giovanni, 53, 54, 58
Bentley, Richard, 101
Berkeley, Bishop George, 145, 198
Bessel, Friedrich Wilhelm, 225
Betelgeuse, 302
Bethe, Hans Albrecht, 289, 299, 300
big bang universe, 201, 246, 266, 267, 272, 318, 329, 338, 346, 347
binary pulsar, 185, 186, 203, 233, 238
binary stars, 184, 240, 307
black hole, 168, 170, 172, 177, 235, 246, 277, 281, 283, 286, 292, 295, 304, 311, 314, 318, 320, 340, 345
black body, 327
black-body spectrum, 273
Blandford, Roger, 235, 282

351

Bohr, Niels, 170, 315, 331
Boltzmann, Ludwig, 325, 326, 329
Bolyai, Janos, 127
Bondi, Herman, 275
Bradley, James, 125
Braginsky, Vladimir, 228
Brahe, Tycho, 35, 41, 44, 80
Brans, Carl H., 176
Brans–Dicke theory, 163, 176, 201, 207, 217
Braun, Carl, 193
brown dwarfs, 288
Bruno, Giordano, 9, 34, 71
Bunsen, Robert, 226
Burbidge, Geoffrey, 280
Buridan, Jean, 52
Buys Ballot, Christophorus H.D., 126

calculus, 13, 83, 86, 115, 116, 120
Cannon, Robert H., 233
Carswell, Robert, 181
Cartesian coordinate system, 56, 97
Cassini, Giovanni Domenico, 124, 249
Cavendish, Henry, 192, 193, 204
centre of gravity, 13, 62, 99
centripetal force, 90, 93, 98, 106, 121, 230, 277, 306
Cepheid variable stars, 182, 260, 262
CERN, 154, 333, 339
Challis, James, 123
Chandrasekhar Limit, 303, 315
Chandrasekhar, Subrahmanyan, 91, 178, 312, 316
China, 1, 51, 346
chronology protection conjecture, 180
Clairaut, Alexis, 119
Clifford, William Kingdom, 173
COBE, 272, 274
conics, 49
constant of gravitation, 191, 192, 329, 339
coordinate geometry, 53
Copernican revolution, 28, 77
Copernicus, Nicolaus, 28, 33, 59, 81, 145, 248
cosmic censorship conjecture, 178
cosmic microwave background radiation, 269, 272
cosmological constant, 182, 255, 256, 294
cosmology, 8, 21, 173, 254, 257, 270, 338
Coulomb, Charles-Augustine, 133
Cowsik, Ramnath, 230
Crab nebula, 186, 305
Critchfield, Charles, 299

Croll, James, 252
Curtis, Heber, 276
Cygnus A, 278
Cygnus X-1, 320

da Vinci, Leonardo, 53, 344
Dalton, John, 325
Darwin, Charles, 297
Darwin, Sir George Howard, 250
Davy, Sir Humphry, 133
de Breteuil, Emilie, 117
de la Condamine, 249
de Maupertuis, Pierre-Louis, 109, 249
De Revolutionibus, 29, 31, 248
de Sitter, Willem, 166, 234, 256, 264
deferent, 16, 17, 21, 46
deflection of a light beam, 166, 212
Democritus, 6
Descartes, René, 55, 59, 95, 104
Dialogue Concerning Two Chief World Systems, 72, 81
Dialogues Concerning Two New Sciences, 63, 74, 81
Dicke, Robert Henry, 162, 176, 201, 202, 228, 270
Dirac, Paul A.M., 154, 196, 345
Doppler effect, 126, 285
Doppler shift, 164, 207, 259, 284
Doppler width, 208
Doppler, Christian, 125
Dreyer, J.L.E., 276
dynamics, 62, 84, 88, 92, 110
Dyson, Sir Frank, 166

Earth, 4, 7, 9, 10, 14, 29, 47, 100, 177, 192, 248, 252, 271, 297, 344, 346
eccentrics, 16, 21
Eckart, Andreas, 283
ecliptic, 14, 16, 17, 243, 252, 253
Eddington, Sir Arthur Stanley, 166, 189, 234, 263, 297, 298, 313, 314
Egypt, 1, 2, 4, 12, 14
Eightfold Way, 333
Einstein, Albert, 55, 96, 108, 142, 143, 144, 146, 155, 157, 158, 162, 167, 168, 184, 254, 313, 330
Einstein–Lorentz transformations, 146, 147, 149
electricity, 131, 134, 135, 173, 334
electromagnetic force, 135, 140, 168, 196, 268, 331

electromagnetism, 135, 136, 154, 174, 266
electrostatic force, 133, 191
electroweak field, 333, 334
electroweak theory, 333
elliptical galaxies, 277, 278
energy, 65, 104, 107, 108, 110, 135, 144, 151, 153, 184, 280, 299
Epicures, 6
epicycles, 16, 17, 46
equant, 16, 46
Equivalence Principle, 100, 156, 163, 175, 204, 223, 330, 340
Eratosthenes, 14
escape speed, 171, 172, 282
ether, 8, 83, 87, 88, 137, 138, 140, 141, 166, 271
Euclid, 12
Eudoxus, 4, 8
Euler, Leonhard, 94, 109
event horizon, 317
exoplanets, 322
expansion of the universe, 169, 255, 269, 293
experimental physics, 61

Fairbank, William M., 233
Far Infrared Absolute Spectrophotometer, 273
Faraday, Michael, 134, 135, 136
Fermat, 110
Fermi, Enrico, 292
Feynman, Richard, 154, 289, 332
field equations, 135, 159, 257
field quanta, 332, 334, 338
field theory, 135, 173, 174, 335
fifth force, 204
First Cause, 25
First Law of Thermodynamics, 108, 326
Fischbach, Ephraim, 204
FitzGerald, George F., 140
FitzGerald contraction, 147
Flamm, Ludwig, 168, 179
Flamsteed, John, 88, 113
flat space, 13
force, 10, 42, 47, 66, 75, 93, 129
Foucault pendulum, 234
Foucault, Jean-Bernard-Léon, 233, 234
Fowler, R.H., 313
Fowler, William F., 280
frame-dragging, 235, 236
frames of reference, 95, 96, 150, 160
free fall, 52, 55, 64, 101, 205
Fresnel, Augustin-Jean, 137
Friedmann, Alexander Alexandrovich, 257, 258

galaxies, 107, 182, 260, 269, 272, 277, 286, 288, 290
Galileo, Galilei, 61, 66, 70, 72, 73, 74, 81, 249
Galle, Johann Gottfried, 123
Gamow, Georgii Antonovich, 264, 290
gauge symmetry, 332, 333, 347
Gauss, Carl Friedrich, 127, 128
Gell-Mann, Murray, 333
general theory of relativity, 144, 158, 178, 254, 317, 330
Genzel, Reinhard, 283
GEO-600, 242
geodetic effect, 234
Georgi, Howard, 334
Giese, Tiedemann, 31
Glashow, Sheldon, 169, 333, 334
global positioning satellites (GPS), 164, 211
gluons, 334
gnomons, 14
Gödel, Kurt, 179
Gold, Thomas, 275, 309
grand unified theory, 334
gravitation, 47, 73, 85, 86, 90, 99, 132, 149, 157, 173, 191, 226, 266, 297, 329, 335, 339
gravitational collapse, 101, 178, 277, 280, 304, 317
gravitational lenses, 180, 181, 182, 291
gravitational radiation, 184, 203, 245, 304, 309, 318
gravitational radius, 173, 285
gravitational redshift, 163, 164, 207, 210
gravitational tidal force, 281, 282
gravitational time dilation, 163, 164
gravitational waves, 184, 187, 238, 239, 246, 304
gravitomagnetic effect, 177, 236, 282
graviton, 175, 266, 336
gravity, 9, 58, 85, 100, 117, 144, 151, 159, 161, 171, 174, 224, 238, 276, 281, 293, 295, 310, 318, 320
Gravity Probe B, 237
gravity shield, 247
Green, Michael, 336
Grossmann, Marcel, 142, 157
Guth, Alan, 268
gyroscope, 177, 233, 234, 236, 238, 252

hadron era, 268
hadrons, 333
Halley, Edmond, 88, 90, 111, 119, 121
Halley's Comet, 90, 103, 119, 171

Hamilton, Sir William Rowan, 110, 120
Hawking, Stephen, 118, 174, 178, 180, 318, 319
Hazard, Cyril, 279
heliocentric model, 21, 35, 37, 41
helium, 232, 237, 241, 265, 269, 292, 298, 299, 339
Heracleides, 4
Herman, Robert, 201, 265, 269
Herschel, William, 122, 276
Hertz, Heinrich, 136
Hertzsprung, Ejnar, 260, 300
Hertzsprung–Russell diagram, 300
Hewish, Anthony, 308
Higgs field, 335
Higgs, Peter, 335
Hilbert, David, 105, 158, 345
Hipparchus, 14, 37
Hipparcos, 217
Hitler, 329
Hooke, Robert, 89
Hoyle, Fred, 174, 266, 270, 275, 280, 346
Hubble constant, 182, 261, 262
Hubble Law, 261, 281
Hubble Space Telescope, 262, 284
Hubble, Edwin Powell, 258, 260, 277
Hulse, Russell, 185, 203, 238
Humason, Milton, 260
Huygens, Christiaan, 83, 84, 85, 104, 165, 312
hydrogen, 192, 265, 279, 299
hyle, 7
Hypatia, 18

Ibn al-Haytham, 21
Ibn Khaldun, 22
Ibn Rushd, 21
ice ages, 252
impetus, 52
India, 1, 20, 346
inertia, 9, 59, 66, 84, 92
inertial frames of reference, 95
inflationary theory, 267
Inquisition, 34, 59, 70, 73
interacting galaxies, 278
interglacial period, 252, 254
International Ultraviolet Explorer, 182
invariable plane, 107, 320
inverse square law, 48, 86, 100, 203, 207, 340
Io, 251
irregular galaxies, 277
Islam, 19

Islamic Spain, 22
island universes, 260

Jansky, Karl, 213
Jeans length, 295
Jeans mass, 295
Jeans, James, 295
Joule, James Prescott, 108
Jupiter, 35, 45, 68, 100, 107, 124, 248, 249

Kaluza, Theodor Franz, 168
Kaluza–Klein theory, 168, 198, 336
Kant, Immanuel, 260, 321
Kepler, Johannes, 41, 42, 44, 47, 50
Kerr, Roy, 319
kinematics, 62, 86, 98
Kirchhoff, Gustav Robert, 226, 327
Kirzhnits, David, 268
Klein, Felix, 105
Klein, Oskar, 168
Kuroda, Kazuaki, 195

LAGEOS, 235
Lagrange point, 121, 306, 319
Lagrange, Joseph-Louis, 110, 120, 121
Laplace, Pierre Simon, 99, 107, 121, 122, 165, 249, 311, 321
Large Electron Positron Collider (LEP), 268, 333, 339
Large Hadron Collider (LHC), 268, 338
Large Magellanic Cloud, 260, 293, 305
large number hypothesis, 196, 198
latitude of forms, 52
Lavoisier, Antoine-Laurent, 122
laws of conservation, 105, 153
laws of motion, 83
laws of planetary motion, 105, 151
le Verrier, Urbain-Jean-Joseph, 123, 132
Leavitt, Henrietta, 260
Lemaître, Georges Edouard, 262, 263, 264
Lense, J., 235
Lense–Thirring effect, 177
Leo I, 262, 291
Leo I local group, 262
libration, 74, 121, 202
light elements, 265
LIGO, 242
Linde, Andre, 268
LISA, 243
Lobachevsky, Nicholas, 127
Locke, John, 112, 116

Lord Kelvin, 140, 297
Lorentz transformation, 141
Lorentz, Hendrik Antoon, 141, 149
Lowell, Percival, 124
lunar laser ranging (LLR), 199, 200, 232
Luther, Gabriel G., 193
Luther, Martin, 30, 33
Lyceum, 5
Lyell, Charles, 297

M31, 260, 284
M33, 260
M42, 276
M87, 284
Mach Number, 197
Mach, Ernst, 197
Mach's principle, 176, 197, 198, 270
MACHO, 291, 292
Magellanic Clouds, 260, 278, 291
magnetism, 131, 133, 135, 136, 331
Main Sequence, 288, 300
MAP satellite, 274, 339
Maric, Mileva, 142, 143, 144
Mariner 6, 221
Mariner 7, 221
Mariner 9, 199
Markarian, B.E., 278
Mars, 45, 46, 124, 199, 218, 222
Maskelyne, Nevil, 192
Maxwell, James Clerk, 108, 135, 136, 137, 141, 173, 193, 326
Mayor, Michel, 322
McKeller, Andrew, 269
mechanics, 13, 33, 62, 63, 67, 91, 98, 104, 109, 110, 223
Mercury, 4, 30, 131, 162, 199, 219, 251
Mesopotamia, 1, 2
Messier, Charles, 275
Michell, Reverend John, 165, 192, 311
Michelson interferometer, 138, 241
Michelson, Albert, 137, 139, 340
Michelson–Morley experiment, 137, 140, 141, 145, 175, 345
Milankovitch Model, 252
Milankovitch, Milutin, 252, 254
Milky Way, 68, 122, 277, 283, 287, 291
Milne, E. A., 313
Minkowski, Hermann, 143, 149, 150
Minkowski, Rudolf, 278
missing mass, 205
Mohammed, 19

momentum, 65, 92, 94, 104, 106, 110, 153, 177, 330
Montagu, Charles, 112
Moon, 1, 4, 42, 47, 68, 74, 199, 200, 224, 232, 251, 279
Morley, Edward, 138, 139
Morris, Michael, 179
Mossbauer effect, 209
Mossbauer, Rudolf, 209
motion, 9, 25, 46, 51, 52, 56, 59, 60, 62, 64, 66, 67, 72, 83, 90, 91, 93, 94, 108, 142, 150, 160
Müller, Johannes, 27

naked singularities, 178
Nambu, Yoichito, 335
Narlikar, Jayant, 174
Nasir al-Din al-Tusi, 21
Neéman, Yuval, 333
negative curvature, 130, 161
neutrino, 269, 292, 305, 310, 339
neutrino astronomy, 306
neutron star, 171, 178, 186, 233, 292, 304, 308
Newton, Isaac, 81, 83, 89, 90, 93, 98, 101, 102, 112, 113, 114, 116, 166, 171, 223, 248, 254, 335
NGC4258, 284
Nobel Prize, 138, 141, 167, 187, 209, 270, 300, 308, 313, 315, 328, 332, 333, 334, 345
Nordström, Gunnar, 168
Nordtvedt, Kenneth, 224, 232
Novikov, Igor, 316
nuclear physics, 164, 264, 298

Obry, Ludwig, 234
Oort, Jan Hendrik, 287
Oppenheimer, J. Robert, 178, 315, 316, 318
Opticks, 82, 113, 114, 116
Orion Nebulae, 84, 276, 302
Ørsted, Hans Christian, 134
Osiander, Andreas, 31, 32
Ostriker, Jeremiah, 288

Pacini, Franco, 309
Paczynski, Bohdan, 291
parallax, 31, 37, 40, 90, 225
parallel axiom, 12, 127, 129
Pauli, Wolfgang, 292
Payne-Gaposchkin, Cecilia, 298
Peebles, James, 270, 271, 288
Penrose, Roger, 178, 281, 318

Penzias, Arno, 201, 270
perihelion advance, 162, 163, 186
Peripatetic, 5, 12
Peuerbach, George, 27
Philolaus, 29, 30
Philoponus, John, 51
Philosophical Transactions, 79
photon, 143, 152, 163, 164, 208, 316, 332
physics, 9, 11, 61, 105, 143, 158, 266, 340
Picard, Jean, 248
Piccolomini, Archbishop Ascanio, 73
Planck length, 329, 337
PLANCK satellite, 274, 339
Planck time, 246, 266, 267
Planck, Max Karl Ernst Ludwig, 326, 327, 328, 330
Planck's constant, 328
Plato, 3, 11, 325
Pleiades, 68
Pluto, 124, 251, 290
Poincaré, Jules-Henri, 95, 141, 148, 149, 155
Pole star, 106, 252
positive curvature, 130, 161, 178, 255
Pound, Robert, 208, 209
Poynting, John Henry, 192
precession, 14, 117, 162, 177, 234, 236, 237, 252
Prime Mover, 8, 11, 18, 46
Principia, 90, 93, 95, 98, 111, 114, 117, 225, 248
principle of least action, 109, 110
principle of least time, 110
principle of relativity, 95, 97, 142, 156
principle of special relativity, 146
probabilistic froth, 329
Procyon, 303
proton–proton cycle, 299
proto-planetary disc, 321
proto-star, 320, 321
proximate cause, 10, 63, 66
Proxima Centauri, 322
Ptolemaic system, 15, 29
Ptolemy, Claudius, 14, 15, 16, 17, 140
Pugh, George E., 233
pulsars, 185, 203, 233, 238, 308
Pythagoras, 2, 43, 57

quanta, 328
quantum chromodynamics, 334
quantum electrodynamics, 154, 332
quantum mechanics, 110, 154, 298, 314, 319, 330, 339, 341

quantum theory of gravity, 175, 330
quarks, 268, 310, 334, 335
quasars, 169, 181, 214, 279
Queloz, Didier, 322
Quinn, Helen, 334
quintessence, 7, 8
Qur'an, 19

radio telescopes, 201, 214
Raman, C.V., 313
Rebka, Glen, 208, 209
red giant, 302, 306
redshift, 164, 182, 257, 259, 269, 279, 295, 317
Refsdal, S., 181
Regiomontanus, 27
Reissner, Hans, 168
resolution of forces, 54
retrograde motion, 4, 16, 29, 219, 344
Rhäticus, George Jochim, 31
Riemann, George, Bernhard, 127, 130, 157, 173
Robertson, Howard Percy, 258
Roche, Edouard, 306
Roche lobe, 306
Roemer, Olaf Christensen, 124, 147
Roosevelt, President Franklin D., 167
rotation of galaxies, 290
Royal Mint, 112
Royal Society, 78, 85, 111, 113, 166
Rubin, Vera, 271, 289, 290
Rudolphine Tables, 42
Russell, Henry Norris, 260, 298, 300
Rutherford, Ernest, 262
Ryle, Martin, 308

Sagan, Carl, 180
Salam, Abdus, 169, 333
Sanduleak–69 202, 305
Schiff, Leonard I., 175, 233, 236
Schmidt, Maarten, 279
Scholasticism, 11
Schönberg, Nicholas, 31
Schonberg–Chandrasekhar limit, 302
Schrödinger, Erwin, 330
Schwarz, John, 336
Schwarzschild radius, 170, 171, 317
Schwarzschild, Karl, 170, 312, 318
Schwinger, Julian, 154, 332
Sciama, Dennis, 271, 316
scientific academies, 77

Scientists' Corner, 114
Second Law of Thermodynamics, 326, 327, 328
self-gravity, 101, 224
Seyfert, Karl, 278
Shapiro, Irwin, 199, 215, 218, 232
Shapley, Harlow, 260, 263, 276
singularity, 178, 258, 264, 275, 317, 329, 341
Sirius, 303
Slipher, Vesto M., 259, 260
Snider, Joseph, 209
Socrates, 2, 3
SOHO, 121, 163
solar system, 15, 29, 38, 98, 121, 202, 290
Sommerfeld, Arnold, 299, 313
space, 9, 13, 83, 94, 96, 127, 135, 144, 145, 149, 151, 160, 174, 197, 266, 341
space-time, 150, 157, 158, 160, 161, 170, 178, 184, 191, 224, 234, 268, 312, 319
special theory of relativity, 143, 144, 160
spiral galaxies, 107, 277, 278
SQUID, 232
Starobinsky, Alexis A., 268
steady-state theory, 275
Steinhardt, Paul, 268
STEP, 230
Stevin, Simon, 54
Stjerneborg, 36
strong equivalence principle, 101, 225, 232
strong nuclear force, 268, 331
Struve, Otto, 315
Sun, 1, 4, 29, 42, 46, 47, 58, 86, 121, 132, 161, 162, 207, 218, 222, 228, 250, 252, 287, 300
sunspots, 70, 72
Supernova 1987A, 293
supernova, 36, 41, 45, 246, 281, 294, 304, 305, 307
superstring theory, 198, 233, 335, 336, 337, 339, 340
symmetry breaking, 268, 333
Szilard, Leo, 167

tachyons, 336
Tartaglia, Niccolo Fontana, 53
tau Boötis, 323
Taylor, Joseph, 185, 203, 238
Taylor, Roger, 323
telescope, 37, 50, 67, 86, 259, 344
Teller, Edward, 167

TEMA, 242
temperature, 202, 254, 266, 269, 273, 300, 302, 319, 321, 326
theory of relativity, 55, 129, 141, 197
Thirring, Hans, 235
Thomson, William, 297
Thorne, Kip, 179
tidal capture binary, 282
time, 9, 83, 94, 95, 141, 144, 145, 148, 149
time machine, 179
time travel, 179
Tomonaga, Shinichiro, 154, 332
torsion balance, 133, 192, 194, 204, 226, 228
Towler, William R., 193
Type I supernova, 307
Type Ib supernova, 304
Type II supernova, 304, 307

ultraviolet catastrophe, 327
uncertainty principle, 330
unified field theory, 168, 169, 335
universal gravitation, 47, 75, 98, 101, 116
universe, 4, 6, 10, 30, 33, 101, 174, 196, 254, 257, 264, 266, 274, 293, 338
Uraniborg, 36
Uranus, 122, 124

van Stockum, J., 179
velocity, 64, 95, 103, 139, 149, 257, 261, 287, 289
Venus, 4, 68, 90, 132, 199, 219, 305, 322
Very Long Baseline Interferometer (VLBI), 216, 237
very massive stars, 292
Viking spacecraft, 199, 222
VIRGO, 242
Virgo cluster, 262
voids, 6
Volkoff, George, 315, 318
Voltaire, 83, 109, 117
von Eötvös, Baron Roland, 157, 175, 204, 225, 227, 228
von Helmholtz, Hermann, 226
von Leibniz, Gottfried Wilhelm, 60, 85, 104, 109, 114, 116, 120
von Soldner, Johann George, 165
Voyager 1, 211
Voyager 2, 211
Vulcan, 132

Walsh, Dennis, 181, 182
wave mechanics, 330
weak equivalence principle, 100, 204
weak nuclear force, 268, 331
Weber Bar, 239, 241
Weber, Joseph, 239, 240
Weber, Wilhelm, 128
Weinberg, Stephen, 169, 333, 334
Weymann, Ray, 181
Wheeler, John Archibald, 180, 311, 316, 345
white dwarf, 208, 224, 262, 295, 303, 307, 313, 314, 315
Wien, Wilhelm, 327
Wigner, Eugene, 167
Wilkins, John, 78
Wilson, Robert, 201, 270
WIMP, 293

world-line, 150, 330
wormholes, 168, 179
Wren, Christopher, 89

X-ray binaries, 168, 281

Young, Thomas, 137
Yurtsever, Ulvi, 179

Zaritsky, Dennis, 291
Zeeman, 141
Zeeman effect, 141
Zel'dovich, Yakov Borisovich, 316
zero curvature, 130, 294
Znajek, Roman, 235, 282
Zweig, George, 333
Zwicky, Fritz, 288, 308, 315